Mechanical Engineering
at the
University of Arkansas

MECHANICAL ENGINEERING

at the

UNIVERSITY OF ARKANSAS,

1874–2004

William Jordan Patty

Phoenix International, Inc.
Fayetteville
2004

Inquiries should be addressed to:
Phoenix International, Inc.
1501 Stubblefield Road
Fayetteville, Arkansas 72703
Phone (479) 521-2204
www.phoenixbase.com

Contents

Acknowledgments

Many people assisted in the making of this written history of the University of Arkansas Department of Mechanical Engineering. Three members of the faculty, Rick Couvillion, Larry Roe, and Bill Schmidt, met with me numerous times to discuss the progress and to suggest possible sources. Rick Couvillion was my contact in the department throughout the process and provided much-needed assistance. Other members of the faculty and staff were equally as generous with their time and assistance. Faculty and alumni interviews also contributed to the success of the project. Both Cecil Cogburn and Ing-Chang Jong provided interesting stories and insights about the people and events that shaped the department. I would also like to thank Mike Martin, who suggested possible leads to investigate based upon his experience with engineering history at the University of Arkansas.

Introduction

Mechanical engineering at the University of Arkansas (originally called Arkansas Industrial University) began like many other departments with faculty members who taught a variety of courses for the first couple of decades. The courses were often called "Mechanic Arts" and taught by superintendents and instructors, few of whom had actual college degrees. By the turn of the twentieth century, though, mechanical engineering, along with other disciplines, had become a distinct department.

The Arkansas Industrial University was founded in 1871 and held its first commencement on June 27, 1872, with Noah Putnam Gates as acting president. As a land-grant college, engineering education was in the early plans of the university. In his inaugural address on July 3, 1874, the university president Albert W. Bishop outlined the plan of the board of trustees for a university containing four colleges with thirteen subordinate departments, including a mechanical engineering department. The outline may well have been the first recorded mention of a mechanical engineering program specifically.[1]

Known as University Hall in the late nineteenth century, Old Main housed mechanical engineering classrooms and labs. (*Cardinal,* 1897)

During those early years, engineering education did not have separate departments, though the board of trustees indicated the programs would eventually be separated. During the 1873–1874 academic year, Gen. N. B. Pearce, professor of mathematics and engineering, supervised the courses in engineering. Col. O. C. Gray, professor of mathematics and civil engineering, succeeded him in 1875, and Gray served until 1879. By 1877, the program had an adjunct professor of civil and mechanical engineering and mathematics, Charles Waite. J. B. Gordon succeeded him in the fall of 1878. Professor Gordon eventually replaced Gray and filled the newly created position of chair of applied mathematics and engineering in 1880, but he passed away during the school year. R. E. Hardiway arrived from Alabama and replaced Gordon for the remainder of the year. J. D. Treadwell succeeded Hardiway for one year before A. V. Lane chaired the program for two years. Lane resigned in 1884. The board of trustees then created the applied mathematics program that included physics, astronomy, and engineering, and Jay Manuel Whitham headed the program. In 1887, Whitham assumed the position of superintendent of mechanic arts and professor of engineering. Other faculty included W. E. Anderson, adjunct professor, as well as four other instructors.[2]

Although the Morrill Act (1862) enabled states such as Arkansas to establish engineering instruction, strong programs did not develop in many parts of the country until the early 1890s. Educators were undecided about the amount of funding that should be directed toward such programs, probably because high demand for mechanical engineers and other technical professionals did not exist in Arkansas and the region. So, the lack of cohesiveness and direction during the early history of engineering at the University of Arkansas was not uncommon at many American universities.[3]

The First Degree and Establishment of the Department

By the late 1880s, the university administration and state lawmakers began to consider the improvements in the mechanical engineering program. Development of the mechanic arts was one of the main objectives of the university, and since good shop facilities were useful in maintaining the university plant as well as for instruction of students, the mechanical engineering area received considerable support earlier than other engineering programs. In 1889 the state legislature appropriated $7,000 for mechanical engineering laboratory equipment, in addition to the $2,500 made available by the board of trustees. At that time, the laboratories were located in the north end of the University Hall (now called Old Main) basement. The 1889 legislature also appropriated $5,000 for a shop building to house the engineering laboratories. By 1894, equipment included a 30-horse-power Corliss steam engine, a 60,000-pound tensile testing machine, and considerable shop equipment.[4] In 1891, the university awarded the first bachelor of mechanical engineering (BME) degree to Mack Martin. Dr. Charles V. Kerr, a mechanical engineer, chaired the engineering department from 1891 until 1896 when George M. Peek succeeded him.[5] Also in 1896, the board of trustees voted to create two departments from what had been one. The mechanical and electrical engineering departments began to function separately after having been grouped together under mechanical arts and engineering.[6]

Chapter 1

The Formative Years, 1897–1923

Birton N. Wilson could be tied to the formative years of the Mechanical Engineering Department more than anyone else since he served in some capacity from 1897 until 1923. His presence as department head provided much-needed stability and leadership. Although the department experienced the change of many of the faculty and staff during the formative years, facilities were consistently upgraded to provide students with the best technical training possible in the state.

Department Heads
George M. Peek, 1897–1898
Charles E. Houghton, 1898–1903
Birton N. Wilson, 1903–1916, 1917–1923 (Experimental Engineering and Drawing)
Frederick G. Baender, 1917–1923 (Heat Power Engineering and Mechanic Arts)

Faculty
1. George Merideth Peek, BME, CEE, University of Virginia; Superintendent of Mechanic Arts and Professor (1897–1898)
2. Birton Neill Wilson, BSME, Georgia School of Technology; ME, University of Michigan; Instructor (1896–1899), Adjunct Professor and Assistant Superintendent of Mechanic Arts (1899–1903), Superintendent of Grounds and Buildings (1899–1903), Professor and Superintendent of Mechanic Arts (1903–1917), Professor of Experimental Engineering and Drawing and Department Head (1917–1923)
3. Mack Martin, BME, ME, University of Arkansas; Assistant Superintendent of Mechanic Arts (1897–1903)
4. Charles Edwin Houghton, AB, MME; Professor and Superintendent of Mechanic Arts (1898–1902)
5. Jubal Early Beavers; Acting Adjunct Professor and Assistant Superintendent of Mechanic Arts (1902–1903)

6. Will Addie Harding; Assistant Superintendent of Mechanic Arts (1902–1903), Machinist (1903–1905)
7. Theodore Charles Treadway, BEE; Adjunct Professor and Assistant Superintendent of Mechanic Arts (1903–1905)
8. Brainard Mitchell Jr., BME, ME, University of Arkansas; Adjunct Professor and Assistant Superintendent of Mechanic Arts (1905–1908), Associate Professor and Assistant Superintendent of Mechanic Arts (1908–1912), Associate Professor (1913–1917)
9. Ellis Blaine Critzer; Instructor in Mechanical Engineering (1905–1907)
10. John Grissom; Engineer (1905–1907)
11. Herman Wakeman Dean; Instructor (1907–1916)
12. W. T. Crippin; Engineer (1907–1916)
13. William Terry Field; BME, University of Arkansas; Adjunct Professor (1908–1909)
14. William Edgar Duckworth; Instructor (1908–1915)
15. Orson Allen Carnahan, BME, University of Illinois; Associate Professor (1912–1913)
16. Claude Bethel; Instructor (1913–1915)
17. Francis Aldrigde Humphreys; Instructor (1915–1916)
18. Frederick Gottlieb Baender, BME, University of Iowa; MME, Cornell University; Professor and Department Head (1916–1917), Professor of Heat Power Engineering and Mechanic Arts and Department Head (1917–1923)
19. John Henry Clouse; Instructor (1916–1917), Instructor in Heat Power Engineering (1917–1919)
20. John Francis Danner; Assistant (1916–1917), Assistant in Heat Power Engineering (1917–1919)
21. James Dinwiddle; Assistant (1916–1917), Instructor in Heat Power Engineering (1917–1923)
22. James M. Hull; Instructor in Heat Power Engineering (1919–1920)
23. James Arthur Jones; Instructor in Heat Power Engineering (1919–1923)
24. Clare A. Poland, BS, University of Kansas; Instructor in Experimental Engineering (1919–1920)

25. Thomas Bartlett Mullin, BS, Queens University; MS, University of Wisconsin; Instructor in Experimental Engineering (1920–1923)
26. Guy B. Irby; Instructor (1920–1922)
27. Jerry E. Stillwell, BSME, ME, University of Kansas; Instructor in Heat Power Engineering (1921–1922)
28. Andrew Jackson Thompson; Instructor in Heat Power Engineering (1921–1923)
29. James Currie Morison; BS, University of Michigan; Instructor in Heat Power Engineering (1922–1923)

First Academic Year as the Mechanical Engineering Department, 1897–1898

When the Mechanical Engineering Department officially began the 1897–1898 academic year as a separate program, students paid $30 per year in tuition, plus around $60 for various expenses such as washing services and uniforms. The university provided housing in two dormitory buildings, the North Dormitory, which included the kitchen facilities, and the South Dormitory, which had the infirmary. However, only male students could live in the dorms. The finances of the university at the time

Buchanan Hall. (*Cardinal,* 1897)

did not allow for a women's dorm, but the university assisted in finding lodging for them in the homes of suitable families. Compared with university housing, boarding with a family cost about $5 more. Students in the dorms paid no rental fee.[1]

The tuition fee could be waived if the student had been selected as a beneficiary from his or her Arkansas county of residence. The board of trustees allowed for one thousand of these students, and each county received a number in proportion to its population. The county judge selected the recipients of these four-year scholarships by examining grammar, spelling, math, and geography skills based on the students' education at that point. The 1897 Catalog stressed that the judges' evaluation should *"note carefully these requirements; otherwise students coming to the University unprepared incur needless expense and go away disappointed and often discouraged."*[2]

There were 220 students enrolled at the Arkansas Industrial University (the name changed to University of Arkansas in 1899) in 1897. Five pursued the BME degree.[3]

Curriculum, 1897

Freshman, 19 credit hours
Math, Plane and Solid Geometry, Elementary Trigonometry, 3 credits
Math, Algebra, 3 credits
Chem, General Inorganic Chemistry, 3 credits
Phys, General Physics, 3 credits
Engl, English, Language, and Literature, 3 credits
ME, Mechanical Drawing, 2 credits
ME, Shopwork, Woodwork, and Founding, 2 credits

Sophomore, 17 credit hours
Math, Plane and Spherical Trigonometry, Determinants, and Analytical Geometry, 5 credits (2 or more courses)
Phys, Electricity and Magnetism, 3 credits
CE, Surveying and Field Practice, 3 credits (2 or more courses)
ME, Instrumental Drawing, 2 credits
CE, Descriptive Geometry, 2 credits
ME, Shopwork with Founding and Forging, 2 credits

Junior, 23 credit hours

Math, Differential and Integral Calculus, 3 credits

Math, Solid Analytic Geometry (optional), 2nd term, 2 credits

ME, Elements of Mechanism, 1st term, 2 credits

ME, Steam Engineering, 2nd term, 3 credits

ME, Statics and Dynamics, Strength of Materials, and

EE, Dynamo Electrical Machinery, 5 credits (2 or more courses)

CE, Masonry Construction, 1st term, 2 credits

ME, Drawing: Machine Design, 2 credits

ME, Mechanical Laboratory, 2 credits

ME, Shopwork with Forging and Machinist Work, 2 credits

Senior, 26 credit hours

ME, Strength of Materials, Hydraulics, Graphics, 1st term, 5 credits (2 or more courses)

ME, Steam Engine Design, Valve Gears, Indicator Practice, 1st term, 3 credits (2 or more courses)

Chem, Technical Chemistry, 3 credits

ME, Locomotive Mechanism, Marine Engines OR French I or German I, 3 credits

ME, Mechanical Refrigeration, Heating and Ventilating, 2nd term, 3 credits (2 or more courses)

ME, Pumping Machinery, Turbines, 2nd term, 3 credits (2 or more courses)

ME, Gas Engines, 2nd term, 2 credits

CE, Engineering Laboratory, 2 credits

ME, Drawing: Steam Engine and Boiler Design, 2 credits

ME, Thesis[4]

After Houghton assumed the leadership role in 1898, course requirements for the BME changed. The hours had been divided out to make course requirements more clear than the 1897 catalog. The freshman and sophomore years remained much the same, but the junior and senior requirements changed considerably. Also, an elective the senior year could be used for a class within the department so the student could focus on a specialized area. Houghton resigned January 1, 1903, and Birton N. Wilson became the acting head but remained until 1916. The options that had been presented to students in the 1898 catalog had disappeared altogether by 1902, resulting in a more uniform and less confusing set of requirements. The department also designated the senior thesis as a second-semester course worth two credits.[5]

Changes in the 1903–1904 catalog under Wilson included the addition of two new mechanical drawing courses, one in lettering and geometrical and the other in perspective and isometric. Also, the department shifted civil and electrical engineering courses from the sophomore year to the junior, removed two hours from the senior elective, and eliminated credit hours from the thesis requirement. During the freshman and sophomore years, students enrolled in algebra for a full semester.

After a year as department head, Wilson implemented a new curriculum in 1903.

Curriculum, 1903

Freshman, 16 credit hours
Math, Solid Geometry, 1st term, Plane Trigonometry, 2nd term, and
Math, Algebra, 5 credits
Engl, English Composition, 3 credits
Phys, Elementary Physics, 3 credits
ME, Mechanical Drawing, 2 credits
ME, Shopwork, 3 credits

Sophomore, 16 credit hours
Math, Determinants, Analytic Geometry, and
Math, Algebra, 5 credits
Chem, General Chemistry, 3 credits
ME, Mechanical Drawing, 2 credits
Phys, Advanced General Physics or a language, 3 credits
ME, Shopwork, 3 credits[6]

Junior, 16 credit hours
Math, Differential and Integral Calculus, 3 credits
ME, Mechanical Laboratory, 2 credits
ME, Mechanics, 4 credits
ME, Machine Design, and

ME, Shopwork, 5 credits
CE, Descriptive Geometry, 1st term, and
EE, Electrical Measurements, 2nd term, 2 credits

Senior, 16 credit hours
ME, Steam Machinery, 3 credits
ME, Mechanical Laboratory (Advanced), 2 credits
ME, Hydraulic Machinery, 1st term, 2 credits
EE, Electrical Machinery, 3 credits
EE, Electrical Laboratory, 2 credits
Elective, 4 credits
ME, Thesis, 2nd term, no credit[7]

When Wilson returned for the 1917–1918 academic year, the department divided into two separate academic units, Experimental Engineering and Drawing, and Heat Power Engineering and Mechanic Arts. Frederick G. Baender chaired Heat Power, which administered the courses required under shopwork, while the drawing courses became the focus of the other department under Wilson, which also included some lab work. Experimental engineering generally concerned how certain mechanisms worked, while heat power engineering courses dealt with materials that contributed to the actual operation of machines.[8]

Despite the separation, students still followed the course requirements for a BME, including a requirement for seniors to complete an economics course in business law. The department reverted back to counting the thesis requirement as two credit hours of the total eighteen during the second semester of the senior year. The university also offered a graduate degree in ME by 1917. The completion of thirty credit hours beyond a BME in addition to a thesis was required. A professional degree could be conferred if a graduate proved to be successful in the chosen field for three years and completed a thesis of original research.[9]

By 1917, mechanical engineering students had to participate in military drill the first two years in accordance with the 1862 Morrill Land Grant Act under which the university had been established. After the first two years, students could choose to continue military service in the Reserve Officers' Training Corps (ROTC) and receive $8 a month in pay. Signing onto the ROTC meant that the student could be called up for duty during college. However, a special provision existed for engineering students called the Engineer Reserve Corps, which enabled engineering students to avoid selective service until the completion of their degree.[10]

Curriculum, 1917

Freshman, 18 hours per term
Fall Semester
Math, College Algebra, 3 credits
Math, Plane Trigonometry, 3 credits
Engl, Rhetoric and English Composition, 3 credits
Chem, General Chemistry Lecture and Laboratory, 4 credits
CE, Drawing, 2 credits
ME, General Shop Practice, 2 credits
Military Art, 1 credit

Spring Semester
Math, Solid Geometry, 3 credits
Math, Analytic Geometry, 3 credits
Engl, Rhetoric and English Composition, 3 credits
Chem, General Chemistry Lecture and Laboratory, 4 credits
CE, Elementary Descriptive Geometry, 2 credits
ME, General Shop Practice, 2 credits
Military Art, 1 credit

Sophomore, 18 hours per term
Fall Semester
Math, Advanced Algebra, 3 credits
Math, Differential and Integral Calculus, 3 credits
Phys, General Physics and Laboratory, 4 credits
ME, Foundry and Pattern Making, 2 credits
ME, Engines and Boilers, 2 credits
ME, Mechanical Drawing, 2 credits
Military Art, 2 credits

Spring Semester
Chem, Metallurgy and Gas Analysis Laboratory, 2 credits

Math, Differential and Integral Calculus, 3 credits
Phys, General Physics and Laboratory, 4 credits
ME, Kinematics, 2 credits
CE, Surveying, 3 credits
ME, Mechanical Drawing, 2 credits
Military Art, 2 credits

Junior, 18 hours per term
Fall Semester
ME, Theoretical Mechanics, 4 credits
ME, Heat Power Engineering, 3 credits
ME, Experimental Engineering, 2 credits
ME, Machine Design, 4 credits
ME, Advanced Shop Practice, 2 credits
Elective, 3 credits

Spring Semester
ME, Strength of Materials, 4 credits
ME, Heat Power Engineering, 3 credits
ME, Experimental Engineering, 2 credits
ME, Machine Design, 4 credits
CE, Hydraulics, 2 credits
Elective, 2 credits

Senior, 18 hours per term
Fall Semester
EE, Electrical Engineering, 3 credits
EE, Elementary Electrical Laboratory, 2 credits
ME, Advanced Experimental Engineering, 2 credits
ME, Mechanical Equipment of Power Plants, 3 credits
ME, Refrigeration, 2 credits
Econ, Business Law, 3 credits
Elective, 3 credits

Spring Semester
EE, Electrical Engineering, 3 credits
EE, Elementary Electrical Laboratory, 2 credits

ME, Heating and Ventilating, 3 credits
ME, Industrial Engineering, 2 credits
ME, Thesis, 2 credits
EE, Electrical Equipment of Power Plants, 3 credits
Elective, 3 credits[11]

When the Mechanical Engineering Department began operating in 1897, the University of Arkansas campus consisted of several classroom and educational buildings other than just University Hall, which housed most of the academic and administration offices throughout the seventy rooms. Engineering and science laboratories, the library, the armory, and the museum also occupied space in University Hall. The Science Hall, built in 1893, contained classrooms and laboratories for the departments of chemistry and physics. The Agricultural Experiment Station consisted of one main building with offices and laboratories and several agricultural structures such as a dairy house and a stock shed. Experiment space for horticultural endeavors included a building for laboratories and a new greenhouse built in 1897. The shops consisted of two buildings, one divided into four areas, the other into two. The main building with four areas contained a foundry, forge shops, a machine shop, and a boiler room. The other building had an office and a storage room for oil and paint.[12]

Mechanical Hall. (*Cardinal*, 1897)

Although University Hall contained laboratory space for mechanical engineering students, the shops, also called Mechanical Hall, provided much of the equipment students needed for experiments. The machine shop included a Westinghouse engine that operated all the equipment in the shops, a large iron planer, a shaper, four lathes of different models of lathes, drill presses, grinding and milling machines, and a Riehle 60,000-pound testing machine. The boiler room included two 75-horsepower boilers, one 40-horsepower boiler, and various instruments such as engine counters and indicators. The forge shop had twelve Buffalo forges with pipes that directed the smoke underground. The woodshop contained an assortment of hand tools and several planers and saws. The foundry featured a Colliau cupola with about three-quarters of a ton of iron per hour capacity, and a brass furnace with 150 pounds of capacity. In addition to the shop equipment, the university also possessed several models and smaller examples of mechanical operations.[13]

The mechanical engineering shops burned three times before experiencing success. The first shops had been in University Hall until 1880 when they were moved to the armory. This structure was destroyed by fire in 1885. The second building burned in 1895 and was replaced by a two-story frame building, Mechanical Hall. The foundry for the shop was built

University Shop Building, 1903. (Mullins Library Special Collections, John H. Reynolds Photo Collection, MSR33, box 1, #17)

Engineering Hall, 1905. (Mullins Library Special Collections, John H. Reynolds Photo Collection, MSR33, box 1, #14)

around a sixteen-inch cupola and a small brass furnace. However, the foundry could not be used because the pattern shop did not have enough tools to operate. Birton N. Wilson secured money for more tools after a $400 deficit was eliminated. The tools he acquired brought the shops up to a satisfactory level of operations, but another fire devastated the shops in 1902. The department soon rebuilt the shops in the winter and secured a twenty-four-inch foundry from Pine Bluff.

The shops found success and actually made money until the foundry was dismantled in 1916, and the pattern shop began to focus on cabinet work. This situation changed about ten years later when the pattern shop began to prepare material once again for a new brass foundry and iron cupola.[14]

When the state made available funds to rebuild the shops after the 1902 fire, the university also secured money to build an engineering hall. The building, completed in 1905 for $25,000, consisted of three floors of classrooms, offices, and laboratories.

Those affiliated with the engineering program celebrated the long-overdue building, but some departments did not receive the same space as others. For example, juniors and seniors in mechanical engineering finally

Engineering Hall Lab, 1905. (Mullins Library Special Collections, photo collection #1090)

received their own drawing room in 1922 when the department secured a room in the southwest corner of the engineering building. Other departments had already had this space for their upperclassmen, while those in mechanical engineering had to share a drawing room with the sophomores.[15]

By the 1920–1921 academic year, a two-fold increase in the number of students led to an almost continuous operation of the mechanical engineering laboratory. To meet this increased demand, the department began to acquire more equipment. In 1922, the mechanical engineering lab received an important addition the department hoped would assist students in thesis work. The equipment included a motor-driven generator with a 550-volt converter and a six-cylinder Mitchell automobile engine. A seventy-two-coil bedspring rheostat of variable resistance comprised the load of the generator, and the turning force was measured directly by small platform scales because the generator was mounted on a dynamotor base. The mounting eliminated the need for determining the output of the motor and generator using a voltmeter and ammeter. The lab also acquired much-needed new equipment such as chisels, hammers, saws, taps, dies, files, and small pipe fittings.[16]

Engineering Lab. (*Razorback*, 1922)

Facilities on and off the campus provided additional opportunities to study applications of mechanical engineering principles. Students examined the new boiler connected to the university power plant, a 250-horsepower Heinie-type inclined drum water tube boiler. Manufactured by the Walsh and Weidner Boiler Company, it could withstand 167 pounds of working steam pressure. Area businesses opened their doors to classes as well. Students in the refrigeration class visited the Fayetteville Ice and Cold Storage Plant. The superintendent discussed with them the energy saved by using different materials in the flooring. Instead of using asphalt made with gravel, they used sand, which made a significant difference in energy costs compared with the old flooring.[17]

In addition to utilizing university facilities to supplement lab equipment, it was not uncommon for faculty members to assist with their maintenance and organization. After strong winds had severely damaged the old stacks that served the boiler, Professor Baender made measurements to extend the breeching of the big stacks over to boiler number four and ordered the materials for the project. Professor Baender also planned to tie the furnace operators' wages to how much coal was used by checking the carbon dioxide recorder that had been recently installed.[18]

The American Society of Mechanical Engineers

The student section of the American Society of Mechanical Engineers (ASME) at the University of Arkansas organized in December 1909, two years after the national society established rules governing student branches. The national society had existed since 1880 when American mechanical engineers could not find any similarities between themselves and the other two established societies for civil and mining engineers. In 1904, a section movement developed to encourage exchange among societies in similar geographical areas. In the 1920s, the Mid-Continent Section, the section Arkansas belonged to, threatened to separate unless the national organization gave petroleum the attention it deserved in the field of machines. The matter eventually dissipated, and the ASME pledged to cover the developments of the petroleum industry more diligently. Initially, the national organization allowed only juniors and seniors to join but changed that policy in 1930 and accepted sophomores and freshmen. The policy change may have resulted from a significant number of national members' inability to pay dues during the depression, so ASME officials decided to offset that by allowing more students to join and, hopefully, pay their dues regularly.[19]

The Reserve Officers' Training Corps

Since 1872, the University of Arkansas curriculum had included some kind

ASME faculty and students. (*Cardinal,* 1913)

of military element as a federal government requirement of land-grant colleges. Military drill required all male students who could to participate, and the university provided uniforms. However, for a variety of reasons, the cadet program struggled to become an integral part of campus life. This situation changed by the early 1900s as the unit grew in size and the

War Department established the Reserve Officers' Training Corps in 1916. After a brief hiatus during the United States's involvement in World War I, the ROTC continued to become popular among both male and female students, especially male students as they received a small paycheck for their participation. During World War II, students readily signed up for ROTC to be placed among the officers when they would be drafted, and the addition of Aerospace Studies during the cold war continued to draw students to ROTC.[20]

The sense of military duty during World War I could be seen on the campus of the university with the establishment of the Student Army Training Corps (SATC) in August of 1918. The United States War Department decided such an organization on university campuses would help aid in the shortage of officers, and the government covered the fees for those students that chose to join the SATC. Perhaps as a result of $30 per month in military pay, most of the male students affiliated themselves with the SATC, and the construction of barracks and heightened security practically turned the campus into a military camp. As more army recruits arrived, patrols around the edge of campus became common, and only those with a pass could gain entrance onto the university grounds. Also, the university switched to a three-term academic year beginning in the fall of 1918. The return to a semester system occurred when the 1925–1926 academic year began.[21]

An influenza epidemic struck the campus and town in the fall of 1918 resulting in the deaths of thirteen students, almost all of them in the SATC. In addition to the widespread illness, the military training consumed much of the students' time despite government assurances to the contrary. When

Military drill. Carnall Hall is in the background. (Mullins Library Special Collections, photo collection #2298)

the SATC disbanded a little over a month after the World War I armistice on November 11, 1918, University of Arkansas president John C. Futrall looked forward to the return of the ROTC, which served a similar purpose but did not require so much of the students' time. The military pressure to streamline courses may have led to some new requirements implemented after the end of the war. In 1919, the College of Engineering required that all freshmen and sophomores enroll and pass the same courses. This practice ceased in 1934 when fall classes commenced, and the college required only that the freshman class complete the same courses.[22]

Despite the rather unpleasant experiences on the University of Arkansas campus during World War I, the military proved to be an important employer for mechanical engineering graduates by the early 1920s. Seniors had the opportunity to sit for a civil service exam for vacancies at Langley Field, Hampton, Virginia, under the National Advisory Committee for Aeronautics. The test consisted largely of questions on general physics, calculus, and electrical, mechanical, and aeronautical engineering. Salaries started at $1,500 to $1,800 a year with increases of $20 a month granted by Congress. The duties included assisting in experimental research by performing calculations, designing apparatus, and compiling reports contributing to the study of aeronautic power plant and aerodynamical problems.[23]

By the early 1920s, employment opportunities existed in many different areas besides the military. Graduates could look for assistant positions on projects such as plant construction. Considerable knowledge of a certain product could be gained as a traveling salesman. Of course, jobs were available working directly for electric or manufacture companies. Although jobs existed, the work included more than just engineering skills. Professor George Stocker, later dean of the College of Engineering, wrote about the concern engineering faculty members had about the poor performance of students in freshman English. He stressed that although engineering students might be tempted to complete only the minimal amount of work needed to pass the course, this would be a regrettable choice later given the importance of good communication skills in many careers.[24]

The University during the Period

While attending the University of Arkansas, students had the opportunity to improve literary skills as well as social skills in general through clubs and societies. Literary societies played an important role in providing students with diversions from classes and the recitations they often involved. Also, restrictions on the social lives of students by the administration and the lack of entertainment in Fayetteville often led to the development of literary organizations to compensate for that void. However, as the administration began to relax rules and as professors required more reading outside of classes, literary societies gradually began to disappear. Also, perhaps due to higher student enrollment, student organizations became more focused and numerous, competing for students' time more than ever before.[25]

The Young Men's and Young Women's Christian Associations (YMCA, YWCA) contributed to student life before and after World War I as the university continued to hire full-time directors for both groups. The organization of events improved when the YMCA moved into a structure that had been built for the SATC by the War Work Council of the National YMCA. Both the YMCA and the YWCA held frequent meetings to plan events for charitable and religious purposes, which became important activities for many students over the years.[26]

Athletics

Although early athletic teams enjoyed only a fraction of the spectators and success as later teams would, sports and fitness began to play an increasingly important role in the early history of the university. The football squad found little success after 1894 when competition with other schools really began. As a mechanical engineering instructor, Birton N. Wilson served as a football trainer for the 1896 team, which illustrates the lack of an athletic staff at the university at that time.

After a series of coaches throughout the early 1900s, Hugo Bezdek coached several winning football teams from 1908 to 1912. It was during Bezdek's tenure that university athletic teams came to be called the Razorbacks after he likened the team's effort to a razorback hog. After Bezdek departed, several different coaches came and went during the 1910s, but the team began to see some success in the early 1920s, first with George W. McLauren then with Francis Schmidt. McLauren led the team to a 3-2-2 record in 1920 and a 5-3-1 record in 1921. When McLauren accepted a position at Cincinnati University, the University of Arkansas hired

B. N. Wilson (*on the far left in the back row*) with the 1896 football team. (*Cardinal,*
1897)

Schmidt, who came on in time to hold spring practice in 1922. Schmidt
led the program to victory, and by 1927, Arkansas boasted an 8-1 record
and a 7-2 record in 1928. After the 1928 season, Texas Christian University
attracted Schmidt to their program, and the University of Arkansas hired
Fred Thomsen. Thomsen would coach several winning teams through the
late 1930s, and led the program to a new height when the 1936 team won
the Southwest Conference Championship outright. After the 1937 team
completed the season with a 6-2-2 record, there would be few wins for the
team over the next four years, and several coaches rotated in and out of the
head coaching position until after World War II.[27]

Although the basketball program did not have a spectacular first year
in 1924, the record improved throughout the 1920s under the coaching
of Schmidt as the team won four consecutive Southwest Conference championships by the end of the 1920s. Due to Schmidt's involvement with
the success of the team, the basketball facility, completed the same year as
the team began competition, came to be known as "Schmidt's Barn." After
a short stint by Coach C. A. "Chuck" Bassett, Glen Rose became coach

and guided the program to several more SWC championships before departing for military duty in 1942. World War II did not slow down the program as the team competed during hostilities and tied for two championships.[28]

The university baseball team competed from 1897 to 1929, with the exception of 1917–1918, until the often harsh weather and low attendance forced the university to cut the sport. The cold weather that usually prevailed until late March also contributed to the baseball teams' lack of practice time and subsequent losing records in the 1920s. The sport did return, though, in 1947.[29]

By the fall of 1919, the university had established athletic facilities on the west side of the campus on Garland Street. The north part of the area included basketball and volleyball courts and a quarter-mile track. The basketball program began the 1924 season in Schmidt's Barn with seating for one thousand spectators and a basement with showers and lockers, and the facility became the center of Razorback athletics. By the next year, a gymnasium for female students had been opened and contained classes, offices, lockers, and showers along with a multipurpose court. Tennis courts had also been constructed by the 1920s, but the university placed them near the dorms. The football and baseball fields occupied the south section along Garland and both fields featured seating areas. The field arrangement had changed by 1926 when the football field was rearranged to run north and south instead of east and west. The baseball field and track then began to occupy the same field north of the football field. By the late 1930s, the athletic facilities began making the shift to the valley below the campus. J. D. McFarland, a 1938 mechanical engineering graduate, worked in the buildings and grounds department and assisted in the construction of the stadium built in the valley.[30]

As an indication of the mutually beneficial relationship, the University of Arkansas and the town of Fayetteville celebrated the school's fiftieth anniversary in June of 1922. The railroads assisted in attracting people to the celebration by reducing rates, and citizens opened their homes to some of the two thousand visitors. The city took advantage of the festivities to make some civic improvements such as paving Dickson Street to the depot and installing streetlights between the university and the town square. The events lasted from June 10 on Saturday until Wednesday when the university held commencement. The celebration also included alumni day on

The University of Arkansas's fiftieth anniversary parade on Dickson Street, 1922.
(Mullins Library Special Collections, Kent R. Brown Collection, Byroade #3)

Monday where such distinguished graduates as United States senator Joe
T. Robinson spoke. President Futrall declared the celebration a success for
attracting alumni back to Fayetteville and encouraging their financial participation in the future growth of the university. Beginning that weekend,
the university financial officer began to oversee an alumni fund for student assistance and general university needs.[31]

Just as the university and Fayetteville had grown in the years since the
founding of the university, so had the mechanical engineering program.
What had been simply a few courses in a general engineering program had
grown into a department with two separate programs. Although the split
department system would not exist beginning in the 1924–1925 academic
year, the Mechanical Engineering Department had indeed made a strong
start that would carry it through the 1920s until the arrival of Professor
Russell G. Paddock in 1931.

Biographical Sketches

George Merideth Peek, BME (1894), CEE (1896), University of
Virginia; Superintendent of Mechanic Arts and Professor
(1897–1898)

The first head of the Mechanical Engineering Department, George Peek had considerable practical experience in the field. He had worked for two railroad yards, the Baltimore and Ohio Railroad in Maryland and the Richmond Locomotive and Machine Works in Virginia. He had also worked in the Baxter Electric Motor Company shops in Baltimore and traveled as an electrician for the company as well. He found employment again in Virginia when he worked in the steam engineering department of the Newport News Shipbuilding and Dry Dock Company in Virginia. He gained teaching experience as an instructor of mechanical and civil engineering for four years while he studied at the University of Virginia.[32]

Birton Neill Wilson, BSME (1896), Georgia School of Technology; ME
(1909), University of Michigan; Instructor (1896–1899), Adjunct
Professor and Assistant Superintendent of Mechanic Arts
(1899–1903), Superintendent of Grounds and Buildings
(1899–1903), Acting Professor and Superintendent of Mechanic
Arts (1903), Professor and Superintendent of Mechanic Arts
(1903–1917), Professor of Experimental Engineering and Drawing
and Department Head (1917–1923)

A native of New York, Birton N. Wilson guided the department for many years after the departure of Charles Houghton. Considering the experience Wilson had working at the university and the practical experience he had before that, the college and the university made a wise decision in selecting him as department head. He worked as a foreman in the Machinist Hall at the Cotton States and Atlanta Exposition in Atlanta, Georgia, and remained there to work as a pattern maker with Glover Machine Works in Marietta. He began his career at the University of Arkansas as the shop instructor in 1896. In 1899 he attained the status of adjunct professor of mechanic arts and superintendent of grounds and buildings. Wilson's position on buildings and grounds might appear odd in the modern era, but at that time, the Mechanical Engineering Department assisted in the maintenance of the university in the days before a physical plant staff. Wilson also served as trainer for the football team when he began employment at the university.[33]

Charles Houghton resigned at the end of 1902, and Wilson served as acting superintendent until the college selected him to be department head in June of 1903. Wilson moved over to a new department in the college called Drawing and Architecture in 1923, which became responsible for the instruction of courses in general engineering drawing. "Boiler Nuts," a nickname based on his initials given to him by students who found his teaching style abrasive, would continue teaching until 1940 when he became emeritus professor of drawing and architecture.[34]

Chapter 2

The 1920s through the Depression, 1923–1941

Beginning of the Paddock Years

During the 1920s, the Mechanical Engineering Department would be unable to retain a department head for more than a few years. The situation with the faculty members would also be unpredictable, but professors such as L. C. Price would begin to make a lasting impression on the department. The construction of the Engineering Hall during the 1920s and the arrival of Professor Russell G. Paddock in 1931 would also be very important for the future of the department.

Department Heads
    Leroy A. Wilson, 1923–1924
    Paul A. Cushman, 1924–1925
    Ernest L. Thearle, 1925–1928
    Clarence H. Kent, 1928–1931
    Russell G. Paddock, 1931–1940

Faculty
1. Leroy Alonzo Wilson, ME, MME, Cornell University; Professor and Department Head (1923–1924)
2. John Coyne Hardgrave, Instructor (1923–1926)
3. Paul Allerton Cushman, SB, Massachusetts Institute of Technology; Professor and Department Head (1924–1925)
4. Joseph Taylor Strate, BSME, University of Wisconsin; MSME, Iowa State University; Instructor (1924–1928), Assistant Professor (1928–1932)
5. Richard C. Milkes, Instructor (1925–1926)
6. Leonard Cassell Price, ME, MME, Cornell University; Research Associate Professor (1926–1942)
7. James Thomas Grimes, Instructor (1926–1928)
8. James C. Harland, Instructor (1926–1946)
9. Robert R. Slaymaker, Instructor (1926–1927)

10. Henry Carl Guhl, BSME, Kansas University; Instructor (1927–1929)
11. Ernest Lathrop Thearle, ME, Cornell University; Professor and Department Head (1925–1928)
12. Clarence Hammond Kent, BSME, Purdue University; MSME, Iowa State University; Professor and Department Head (1928–1930)
13. Frederick Charles Werber, Instructor (1928–1947)
14. Alester Garden Holmes Jr., BSME, Clemson Agricultural College; MSME, Cornell University; Instructor (1929–1938)
15. Marshall Elmer Farris, BSME, Purdue University; MSME, University of Texas; Acting Professor and Department Head (1930–1931)
16. Russell Gurney Paddock, BSME, MSME, Purdue University; Professor and Department Head (1931–1958)
17. Clinton William Janes, BEE, University of Minnesota; Instructor (1937–1940)
18. James Gordon Gleason, BS, Alabama Polytechnic Institute; MSME, University of Arkansas; Instructor (1940–1943), Assistant Professor (1943–1945), Associate Professor (1945–1954), Professor (1954–1984)

The Mechanical Engineering Department in the 1920s experienced instability because department heads did not remain for more than a few years. Other than the department head, few faculty members ranked higher than instructor during the 1920s. F. G. Baender resigned in the spring of 1923 and accepted a position with the Hays Carbon Dioxide Recorder Company. Leroy A. Wilson replaced Baender, as well as B. N. Wilson, as head of the reunified Mechanical Engineering Department. Wilson had previously taught at Oklahoma A&M and served during World War I as a 1st lieutenant in the air service in charge of the engine department of the Army School in Urbana, Illinois. Paul A. Cushman arrived from Brooklyn Polytechnic and served as head during the 1924–1925 academic year. Ernest L. Thearle then accepted the position as department head. Thearle left the university in 1928, and Clarence H. Kent arrived to take his place. He had held positions with Westinghouse, United States Army Corps of Engineers, and the Erie Railroad. Kent did not return for the 1931–1932 school year after taking a leave of absence the year before. After a brief period with the acting department head Marshall E. Farris, Russell G. Paddock accepted the position as department head in 1931 and remained for many years. The faculty turnover rate decreased after Paddock arrived, which could be attributed to his leadership as well as the waning effects of the depression.[1]

Although Leroy Wilson was department head for only one year, he oversaw the changes that resulted from departmental reunification. The requirements were adjusted within the college so that all freshmen and sophomores enrolled in the same classes. Also, the university had changed to a three-term academic year. The total number of credits for each term under a full load was eighteen. Shopwork had been divided into separate courses. Another change made for the 1923–1924 academic year required that seniors take experimental engineering for the entire year rather than just one term. Along with some other adjustments, the department hoped students would find the coursework more challenging so that they would be better prepared upon graduation.[2]

Curriculum, 1923

Freshman, 18 credit hours per term
Phys, General Physics, 12 credits
Engl, Rhetoric and Composition, 9 credits
Math, College Algebra, 15 credits
D&A, Mechanical Drawing, 6 credits
ME, Woodwork, 6 credits
MA, Military Art, 3 credits
Math, Solid Geometry, 3 credits

Sophomore, 18 credit hours per term
Math, Differential and Integral Calculus, 15 credits
Chem, General Chemistry, 15 credits
D&A, Elementary Mechanical Drawing, 6 credits
CE, Elementary Surveying, and
CE, Field Practice, 15 credits
ME, Elements of Mechanical Engineering, and
ME, Elementary Mechanical Lab, 15 credits

EE, Elements of Electrical Engineering, and
EE, Elementary Electrical Lab, 15 credits
MA, Military Art, 3 credits[3]

Junior, 18 credit hours per term
ME, Mechanics, 6 credits
ME, Strength of Materials, 6 credits
ME, Heat Power Engineering, 9 credits
ME, Mechanical Engineering Lab, 6 credits
EE, Principles of Electrical Engineering, 9 credits
EE, Electrical Engineering Lab, 3 credits
ME, Elementary Machine Design, 6 credits
Elective, 9 credits

Senior, 18 credit hours per term
ME, Mechanical Equipment of Power Plants, 3 credits (fall)
EE, Electrical Equipment of Power Plants, 3 credits (winter)
ME, Technical Specifications, 3 credits (spring)
CE, Hydraulics, 4 credits (spring)
Econ, Commercial Law, 4 credits (fall)
Econ, Industrial Management, 4 credits (winter)
ME, Heating and Ventilation, 6 credits (fall and winter)
ME, Advanced Machine Design, 6 credits
ME, Advanced Mechanical Lab, 6 credits
ME, Thesis, 3 credits
Elective, 12 credits[4]

Even though no single figure served as department head throughout the 1920s, changes occurred in the requirements. The department introduced two new courses for the 1925–1926 academic year in metallography and microscopy laboratory, which counted as two-hour courses each. Beginning that same year, engineering students could choose whether or not to complete a senior thesis. Before this year, it had been a requirement at least since mechanical engineering became its own department. The short course in mechanical engineering also did not appear in the 1925–1926 catalog. This had been available since the early years of the department. In the fall of 1925, the university returned to a semester system with fall and spring terms. By 1927, the degree offered in mechanical engineering came to be known as a bachelor of science in mechanical engineering (BSME) rather than a BME. Beginning in the fall of 1929, the College of Engineering established a general engineering department with courses taught by Dean William N. Gladson. The department offered required orientation courses for freshmen and sophomores, which consisted of eight lectures by Gladson over the length of both the freshman and sophomore years.[5]

By the 1940–1941 school year on the eve of the United States's entry into World War II, the college had begun to allow the separate departments to choose their own coursework for the sophomore year. General engineering added a contract law course and dropped the sophomore orientation lecture series. The Mechanical Engineering Department allowed flexibility for students in shop classes. Although the catalog listed the courses as three hours a week, students could reduce the hours and take more than one in a single semester. Also, the department began offering an aeronautical option "to prepare for the more specialized field of aeronautical engineering." Students in their senior year could substitute up to six hours of aeronautical related coursework for their elective courses and some of the required courses. The students still received a BSME, however. One of the requirements of membership in ASME stipulated students enroll in one seminar their junior and senior year and produce a paper to be presented to the organization.[6]

Curriculum, 1941

Freshman
Fall, 17 credit hours
Chem, General Chemistry, 4 credits
Engl, Rhetoric and Composition, 3 credits
Math, College Algebra, 3 credits
Math, Plane Trigonometry, 3 credits
D&A, Mechanical Drawing, 2 credits
ME, Elementary Wood Shop and Foundry, 1 credit
MA, Military Art, 1 credit
GE, Freshman Orientation, no credit

Spring, 18 credit hours
Chem, General Chemistry, 4 credits
Engl, Rhetoric and Composition, 3 credits
Math, Analytic Geometry, 5 credits
Phys, Engineering Physics-Mechanics, 2 credits
D&A, Descriptive Geometry, 2 credits
ME, Elementary Forge and Machine Shop, 1 credit
MA, Military Art, 1 credit
GE, Freshman Orientation, no credit[7]

Sophomore
Fall, 18 credit hours
Phys, General Physics, 4 credits
Math, Differential and Integral Calculus, 4 credits
D&A, Mechanical Drawing, 2 credits
Econ, Principles of Economics, 3 credits
Engl, Composition, 2 credits
ME, Pattern Shop and Machine Shop, 2 credits
MA, Military Art, 1 credit

Spring, 19 credit hours
Phys, General Physics, 4 credits
Math, Differential and Integral Calculus, 4 credits
D&A, Mechanical Drawing, 2 credits
Econ, Principles of Economics, 3 credits
Engl, Public Speaking, 2 credits
CE, Surveying Field Practice and Elementary Surveying, 3 credits
MA, Military Art, 1 credit

Junior
Fall, 17 credit hours
ME, Mechanics, 5 credits
ME, Thermodynamics, 3 credits
ME, Mechanism, 3 credits
ME, Mechanical Lab, 2 credits
EE, Electrical Engineering, 3 credits
Elective, 1 credit

Spring, 18 credit hours
ME, Mechanics, 5 credits
ME, Thermodynamics, 3 credits
ME, Kinematics, 2 credits
ME, Mechanical Lab, 2 credits
ME, Seminar, 1 credit
EE, Electrical Engineering, 3 credits
EE, Electrical Engineering Lab, 2 credits

Senior
Fall, 18 credit hours
ME, Steam Power Plants, 3 credits
ME, Advanced Mechanical Lab, 2 credits
ME, Machine Design, 3 credits
GE, Law for Engineers and Engineering Contracts, 3 credits
CE, Hydraulics, 4 credits
Elective, 3 credits

Spring, 18 credit hours
ME, Internal Combustion Engines, 3 credits
ME, Seminar, 1 credit
ME, Metals, 3 credits
ME, Advanced Mechanical Lab, 2 credits
ME, Machine Design, 2 credits
Mgmt, Industrial Management, 3 credits
CE, Materials Lab, 1 credit
Elective, 3 credits[8]

As student enrollment increased in the mid-1920s, the mechanical engineering lab continued to acquire equipment. By December of 1924, the department added a Willy's Knight sleeve-valve motor, a dynamometer, and a model refrigeration plant. About a month later, the lab also received several motors for test block work. However, space became an issue in the lab with the additional equipment. The department looked forward to a new facility along with the rest of the college, and plans began to materialize for the new engineering building in the fall of 1925. The new building had been prioritized by the early 1920s as the freshmen enrollment climbed and retention rates improved in the college. Classrooms and offices were planned for the three floors and labs would occupy the basement, including one for mechanical engineering in the southwest corner. The Arkansas legislature passed appropriations for the building at the same time they directed money for a new agriculture building. The two buildings would be the first new academic structures constructed in twenty years.

After learning the building received funding, the department began to update its equipment in preparation for the new lab. Two pieces of equipment, a turbo generator from General Electric and a steam engine from Moore Steam Turbine Company, arrived in the fall of 1926. The turbo generator included special features that allowed for manipulations in lab experiments. The steam engine featured a crosshead, which the Moore Company patented. The department decided to acquire another piece of equipment and bought a Fitchburg piston valve uniflow steam engine made by the FSE Company of Fitchburg, Massachusetts. The 40-horsepower engine had a water rate of 20.5 pounds of steam per horsepower hour, and it was designed for lab use so that the points of compression, admission, and release could be varied by the operator. Professors Ernest Thearle and Joseph Strate received much of the praise for the acquisition of the new equipment.[9]

More equipment arrived after completion of the lab. The department decided to add another engine specifically for lab use, an eight-by-ten Ridgeway simple steam engine. The engine featured two independent governors to study brake horsepower and valves. In order to expand the capacity for comparison study, the department ordered a Sullivan two-stage air compressor with inter-cooler, which was

Mechanical Engineering Lab. (*Razorback*, 1934)

directly connected to a 60-horsepower Fairbanks-Morse type Y diesel engine of the solid injection type. The compressor was later hooked up to the main steam line so that the steam equipment could also be run by compressed air. One improvement in the new lab assisted in the installation of the new equipment. A Telepher carriage system greatly assisted with the movement of very heavy materials from the rear door to the labs in the basement.[10]

The Mechanical Engineering Department began the fall semester of 1927 in the new engineering building along with the civil and electrical engineering departments, the Department of Drawing and Architecture, and the Engineering Experiment Station. The top floor of the building housed drafting rooms, offices, classrooms, and a darkroom. The second floor contained offices and classrooms as well, but also a library and an auditorium. The basement, the first floor, contained laboratories for the three engineering departments and the Engineering Experiment Station.[11]

As proof of the benefits in having a well-equipped laboratory, mechanical engineering students had an impressive display during the open house that occurred on Engineer's Day in 1927. In addition to the arrangement of refrigeration equipment and boiler feed pumps for the public to view how they worked, students displayed machines designed and built in the

Engineering Hall viewed from Dickson Street, ca. 1930. (Mullins Library Special Collections, photo collection #1063)

labs. These included Little Julius, the world's smallest working steam engine, and the Purdy Uniflow Engine designed by graduate Russell Purdy, which students claimed was the only poppet valve uniflow engine. By the fall of 1934, the decision had been made to separate the festivities of Engineer's Day from the exhibit displays. Each department would now display their projects in April on the same day of the University High School Meet. Officially, the change ensured a larger audience to enjoy the projects and to showcase the accomplishments of the College of Engineering. Another reason for the change probably addressed faculty concerns that students did not have enough time to prepare projects for display in March.[12]

The movement of some equipment from the shops to the new lab

Engineering Day Shop Display. (*Razorback*, 1926)

proved to be a wise decision as fire once again severely damaged the shops on May 4, 1928. Starting in the foundry room, the fire caused $38,000 in damage to the building itself. Fortunately, insurance covered $30,000, but the uninsured equipment from the machine lab had to be replaced at a cost of around $20,000. Among sentimental losses were the Bucking Ford and the trolley car, both used annually during Engineer's Week.[13]

The despair over the loss of the shops dissipated as the new equipment arrived. Since most of the old equipment had been acquired before World War I, the replacements were quite a bit more advanced. Two portable, inexpensive structures for the shops arrived from the Tucson Steel Company and required few modifications. One building contained the pattern and machine shop, and the other held the forge and foundry. Among the new features in the machine shop were 12 nine-inch South Bend lathes. Also, the shop acquired a large, twenty-one-inch South Bend lathe and a large Grand Rapids universal grinding machine. A low price allowed for the purchase of the thirty-inch Gisholt vertical boring and turning mill, a unique piece of equipment few universities could afford. The pattern shop received individually driven equipment and work-benches, and the foundry received a twenty-one-inch crucible furnace and a thirty-two-inch Whiting cupola. The Buffalo Forge Company supplied the forge shop equipment.[14]

By the fall of 1928, the department had a record enrollment and began offering a course in aerodynamics for the first time in the spring. Professor Strate taught the new class. The course served as an introduction covering elementary dynamics, types of planes, the operations of different parts, and the engines used. The development of an aeronautics course eventually led to experiments with the material discussed in classes. Professor Price and ASME section members undertook a project to build a gasoline engine that airplanes commonly used at the time. One of the problems encountered was the lack of air to cool the engine since the airflow that typically cooled the engine during flight could not be recreated. As a result, the engine could not run for long periods of time.[15]

Another group of students and Price pursued a project to build another airplane engine early in 1933 in the mechanical engineering lab. They decided to build the engine around a two-throw crankshaft that had been sitting in the mechanical engineering drafting room for some time. Since low cost along with good design was the main consideration, the students visited junkyards to acquire parts from both automobiles and motorcycles, and they found some parts around the lab. The engine was completed in February of 1934. It turned the propeller at 2,000 rpm and measured twenty-six inches wide, sixteen inches high, and twenty-five inches long minus the gas tank. The engine weighed 160 pounds and included a car-buretor, an ignition system, and a propeller hub. The horizontal-oppose design could be found in many small planes at the time, and the dimensions conformed to those in Continental A-40 engines.[16]

Events outside of the labs illustrated the need for trained engineers in

Arkansas. An optional event, the inspection trip, nearly turned deadly for junior mechanical engineering students in the spring of 1929. Professor Kent, the department head at the time, wrote an article titled "Caustic Embrittlement in Steam Boilers" in the November 1929 *Arkansas Engineer* to address what had happened in the power plant of the Crossett Lumber Company in Crossett, Arkansas. A new 1,000-horsepower water tube boiler exploded, killing one man and injuring four others. The topic was especially relevant to engineering seniors who had been on their junior field trip there the day before the explosion.[17]

Students experiment with an airplane engine. (*Razorback*, 1939)

The ASME continued to plan activities under financial constraints in the early 1930s. In the spring of 1933, the group, along with Professor Paddock and his wife, went to a lime kiln for a wiener roast and a "wet party." The first meeting held in the fall of 1933 featured talks on "What Is Wrong with the American Automobile" by William Hosford and "Steam Refrigeration and Its Application to Air-Conditioning" by Phil Herget. The second meeting in October included the testing of a Wakesha bus engine in the mechanical engineering lab. The Great Depression did affect membership, however. Ernest Hartford, national ASME assistant secretary, visited the student chapter along with Harold Adkisson, assistant secretary of the midcontinent section, in January of 1935 to discuss reducing dues and initiation costs to attract more students. The strategy appeared to have worked by the fall of 1937 when the chapter had a record number of twenty-eight members.[18]

During a meeting in the fall of 1936, the ASME section proposed to transfer from the South to the North Central division. Many of the students and professors in schools in the university's division often discussed

Professors Price and Paddock with ASME students. Price is in the middle row, second from the left, and Paddock is in the middle row, fourth from the left. (*Razorback*, 1940)

problems of oil production during regional meetings due to a majority being located in the oil belt. The North Central division schools appeared to have more in common with the University of Arkansas concerning mechanical engineering topics due to geographical similarities. The ASME headquarters did not accept the proposal because the national secretary deemed a transfer inadvisable.[19]

Delta Psi, the local engineering fraternity, began its affiliation with Theta Tau, the leading national engineering fraternity, as the Upsilon Chapter on April 7, 1928. A mechanical engineering student who attended the university in the late 1940s recalled his experiences with Theta Tau: "Through my good junior class friends, I became a pledge of Theta Tau. This required the fabrication of a wooden gear with all the proper dimensions, properly decorated as a Theta Tau Gear. It was also necessary to have all members sign my gear, which was usually finished by one or more swats on the rear as an exclamation point. As a part of the preliminaries, all pledges had to perform duties around the chapter house such as cleaning and repairing as instructed by members. If not done to their satisfaction there would be an additional 'Grab the Ankles' for a swat with the pledge paddle. I became a member of Theta Tau and participated in all activities. It was very beneficial to my career in college and all future activities."[20]

Although Theta Tau did not have a house, another group provided rooms for engineering students. By the end of the 1930s, the Engineer Cooperative Housing Organization (ECHO) had been established to provide an inexpensive alternative to university lodging. The group charged rent of $16.50 a month in the former 4-H House. A mechanical engineering student during that time recalled that he paid only $10.50 a month since he served as the treasurer of the co-op. Those who lived there in the late 1930s kept the place in reasonable enough condition to even host the university president, J. William Fulbright, and his wife for dinner.[21]

By the 1931–1932 academic year, the General Engineering Society (GES) constitution, written in 1922, required an update due to changes in the college. The language and some of the topics appeared to be quite dated, so students proposed a new constitution. The revisions included the elimination of passages concerning the engineer's parade and federal students. Additions included establishing the *Arkansas Engineer* as the official organ on the college and including the chemical engineering department as part of the college. The closer affiliation between the GES and the *Arkansas Engineer* led to a decision by the GES to adjust the publication of the periodical so that it might be considered for membership in the Engineering College Magazine Association (ECMA). One of the major changes involved the creation of a permanent board made up of engineering Professors William B. Stelzner and William R. Spencer and Dean William Gladson.[22]

Due to the unstable economic times of the early 1930s, the *Arkansas Engineer* and the agricultural college publication, the *Agriculturalist*, were put under the supervision of the Board of Student Publications so the two periodicals could avoid posting large debts. The *Arkansas Engineer* proposed using a low-quality paper and eliminating several pages from each issue to avoid the suspension threatened by the board. Publication had already been suspended once before during the 1930–1931 academic year.[23]

Eventually, the Engineering Council replaced the GES in 1941 after the U of A hosted an ECMA meeting. After meeting with students from

In this view of the south side of campus, the new engineering hall is on the left, and
the old engineering hall is on the right behind the tower of Old Main, ca. 1930.
(Mullins Library Special Collections, photo collection #938)

other campuses, Willis Dortch, the editor of the *Arkansas Engineer* at the
time, discovered that Engineering Councils had replaced the GES on many
campuses. Apparently, business progressed much easier after the establishment of a council, so student leaders drew up a constitution to replace
the GES on the University of Arkansas campus. After some difficulty, the
constitution passed. Another group developed several years later when students initiated the Arkansas Engineering Society in the fall of 1950 to
develop better contact among the individual engineering societies.[24]

The GES and the Engineering Council both assisted in the festivities
surrounding Engineer's Day, a celebration that originated at the University
of Missouri in 1903. After students in Missouri determined a link between
St. Patrick and the engineering profession, March 17 was designated as a
day of celebration. Students at the University of Arkansas began to hold
their own version in 1909. Over the years, the college acquired several
objects that would only be brought out annually such as an actual piece
of the Blarney stone, the Toonerville Trolley, and the bucking Ford.

By the 1930s, agriculture students had their own day as well, and both
groups of students would pull pranks on the other on their respective days.
On Engineer's Day in 1934, students attempted to paint a shamrock, many

The Bucking Ford. (*Razorback*, 1925)

of which appeared on campus annually around March 17, on the agriculture building, but agriculture students dumped cow chips on them from the roof. On the agriculture day, those students attempted to cause mischief in Engineering Hall, but engineering students were waiting and blasted the intruders with a fire hose. The next year, Engineer's Day came under scrutiny after a rock-throwing incident blamed on the engineering students left many windows in the agricultural building shattered. Because of the incident, the university administration considered revoking festivity privileges for the students in the College of Engineering.[25]

By the end of the 1930s, job prospects for mechanical engineering graduates remained stable. Graduates could find jobs in the manufacturing sector because industry searched for educated employees to anticipate problems before they occurred so money could be saved by eliminating unnecessary steps. Drafting was another option, but wages tended to be insufficient because many draftsmen had no formal education and worked for low wages. Ownership of a business would be perhaps the most difficult as it required a large capital investment. Although engineering graduates had been limited by the depression to federal government jobs, the private sector began to hire more of them by the mid- to late 1930s. As more and more private industries began normal operations once again, companies needed equipment repairs after years of dormancy and neglect.[26]

Mechanical engineering student Billy Lewis was chosen as St. Pat in 1937, and Frances Rossner was selected to be St. Patricia. (*Razorback*, 1937)

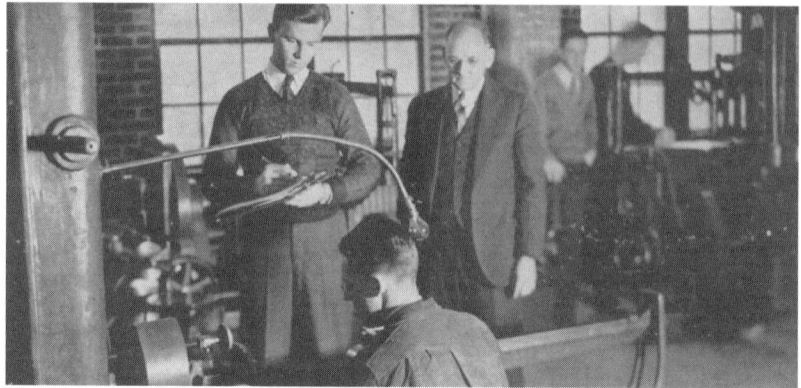

The instructor observes students in the Mechanical Engineering Lab. (*Razorback*, 1937)

Finding a job with a good salary would be especially important living away from the inexpensive housing in Fayetteville. George Doerries, a mechanical engineering graduate, successfully found employment in the early 1940s but discovered that the cost of living changed after leaving Arkansas. According to Doerries, living with his wife in a three-room apartment in New Jersey was a rude awakening compared with what he had paid to live with three roommates in one room while attending the University of Arkansas.[27]

The growth Fayetteville experienced in the 1920s helped to stave off some of the more devastating effects of the depression. The diverse agriculture in surrounding Washington County played an important role in such areas as strawberry and poultry production, and factories to handle the packing and canning of such goods provided stable employment. The town began to market itself as a tourist destination with the completion of the Western Methodist Assembly on Mount Sequoyah in 1923. Business leaders provided financial support for the construction of roads and utilities for the convention site as they anticipated a source of revenue for the summer months when many students departed until fall. Travelers could also find ample rooms on Dickson Street at the OK Hotel, the Shady Lane, the Midway, and the Winkleman. If travelers stepped into the Majestic Café run by George S. Pappas and his brother, they would more than likely have found themselves among many university faculty and students in the popular establishment near the campus. A new sector of the business community developed as more citizens became automobile owners and required gas and service for them. By the end of the 1920s, thirty gas and service stations had opened to vie for the business of approximately one thousand automobiles in the area. Drake Field opened in 1920 and serviced private aircraft along with military planes that used the field as part of its training. The depression did slow down this economic vitality, but none of the banks closed their doors, and, as in many other towns across the country, the Works Progress Administration provided employment working on the town's drainage system and roads.[28]

The Mechanical Engineering Department successfully weathered the depression as well. In fact, faculty retention improved over the course of the 1930s to give the department a place to begin to establish a solid core of professors. The new Engineering Hall no doubt assisted in this endeavor, as well as the leadership of Professor Paddock, who would continue to attract well-qualified faculty members until his retirement in 1958.

East side of the downtown square in Fayetteville, 1920s. (Mullins Library Special Collections, Kent R. Brown Collection, Gresham #4)

Accreditation

In 1924, the college received good news when the New York State Department of Education gave the college its highest approval rating, a tough standard for many engineering colleges to obtain.[29] The Engineering

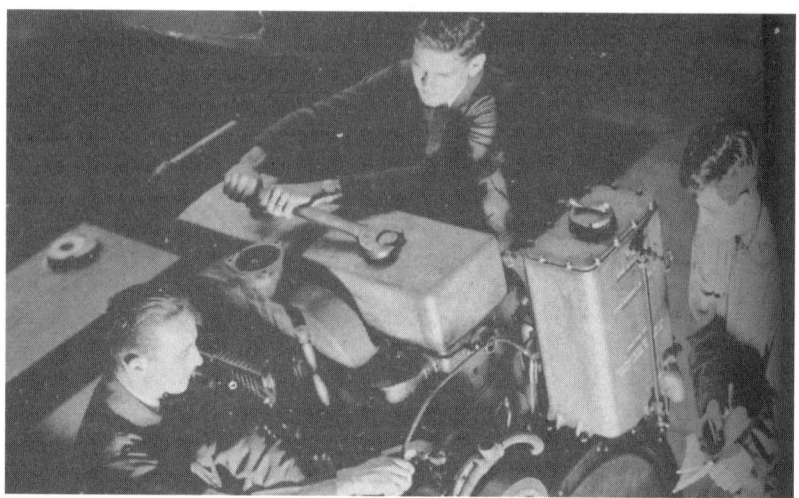

Students in the Mechanical Engineering Lab. (*Razorback*, 1937)

Council for Professional Development, now the Accreditation Board for Engineering and Technology (ABET), began accrediting programs in 1936. The Mechanical Engineering Department applied for and was granted accreditation soon after and has been continuously accredited since that time.

Biographical Sketches

Leonard Cassell Price, ME (1925), Cornell University; Research Associate Professor (1926–1942)

Professor L. C. Price taught Mechanisms and Dynamics of Machinery, among other courses. In addition to teaching duties, Price worked with many students over the years to build engines for the purpose of experimentation. Perhaps his greatest legacy would be the engine he and a class built toward the end of his time at the University of Arkansas. He proposed for the advanced students to build an engine that would be ideal for lab exercises. The students put together an engine with a motorcycle piston, a connecting rod, and a crankshaft. The result allowed for multiple tests to be done. Students could change the spark and read where the spark was firing whenever it was changed. Also, the valve timing and the fuel mixture could be changed while the engine was running. Since the engine was such a welcomed addition to the lab, students named it Elsie after "L.C." Price.[30]

Although he worked well with students, he expected intelligent input rather than statements without meaning. To this effect, he had a sign over the blackboard that read: "It is better to be silent and thought a fool than speak and remove the doubt." A student in the early 1940s "kind of read [the sign] out loud, and he said, 'Okay, I do not need to say anymore.' Everybody laughed up a great storm, and my pride went down into my shoes." Price left the U of A after spring 1942 to accept a position at Michigan State College in Lansing, Michigan. According to the *Arkansas Engineer* in 1942, "He has always been willing to help any student whether he was at the bottom of the class or the top. It will be a big task to find anyone who can even approach his ability; and that empty feeling that will be left in the hearts of the Mechanicals will never be filled."[31]

Russell Gurney Paddock, BSME (1920), MSME (1923), Purdue University; Professor and Department Head (1931–1958)

Professor Russell G. Paddock was born in 1893 and attended Marion College before World War I. He received his BSME from Purdue University in 1920 after serving in the army as a second lieutenant in the field artillery between his freshman and sophomore years. He was a member of both Pi Tau Sigma and Tau Beta Pi honor societies and the Acacia fraternity while at Purdue. After graduation, he accepted a job as chief draftsman for the Indiana Lamp Company in Connersville, Indiana. He returned to Purdue in 1921 as an instructor in mechanical engineering, received an MSME degree in 1923, and accepted a position at the Consumers Power Company in Jackson, Michigan, to test the efficiency of the company's steam plants. After he left Consumers in 1930, he worked for Sargent and Lundy in Chicago surveying industrial power plants before accepting the department head position in 1931.[32]

Courses that Paddock taught included Mechanics, Thermodynamics,

Russell G. Paddock, ca. 1935. (Mullins Library Special Collections, photo collection #644)

Steam Power Plants, Kinematics, and Metals. According to a former mechanical engineering student in the late 1940s, Paddock would give tests in mechanism and kinematics "that would curl your hair." A student from the late 1950s recalled Paddock as "a bear of a man. He looked at you like he really wanted you to graduate, and he worked hard to see that it happened. He graded fifty percent on knowing his material and fifty percent on getting the correct answer. He used a circular slide rule that had more accuracy than our Post Versa-log models swinging from our hips. His final for my heat power ME option was only one question, but it had well more than a dozen sub-parts, each depending on the answer to a preceding sub-part to work a next sub-part. I know that if a one missed a correct answer, Paddock would substitute the incorrect answer to see if later calculations were correct. His door was always open, and he encouraged his students to get help if they needed it. For many of his students, it was the first time to be out on their own, and his support was greatly needed and later appreciated."[33]

Paddock lived in Fayetteville with his wife, Irene, after retiring in 1958. He passed away in 1978. His impact on the Mechanical Engineering Department was described in his eulogy by a quote from Elton Trueblood. Russell Paddock was a man who "planted shade trees under which he knew full well he would never sit."[34]

James Gordon Gleason, BS (1938), Alabama Polytechnic Institute; Instructor (1940–1943), Assistant Professor (1943–1945), Associate Professor (1945–1954), Professor (1954–1984), Professor Emeritus (1984–2000)

James G. Gleason joined the Mechanical Engineering Department in 1940 after receiving a year of training at the New York State Merchant Marine Academy, located at Fort Schuyler in the Bronx. He also worked as an instructor in a mechanical engineering lab and for Mercury Aircraft, Inc. He received his engineering degree from Alabama Polytechnic Institute in Aeronautical Engineering in 1938. He remained there to instruct Thermodynamics and Mechanical Lab for seniors and juniors and oversaw the Civil Aeronautics Authority. He came to the University of Arkansas in the fall of 1940 to head aeronautical studies including the flight training program in addition to teaching Mechanics and Machine Design.[35]

Gleason played an important role in the department, especially concerning the arrangement of the labs in both the old and the new buildings. In the old building, he directed the rearrangement of equipment around Elsie to improve quality of the experiments for the students. When the department moved to the new Mechanical Engineering Building in 1965, Gleason worked to bring in new equipment since many of the

James G. Gleason, ca. 1981. (Mechanical Engineering collection)

engines were discarded or sold. According to Cecil Cogburn, "He basically built that lab." One of the improvements over the old lab was to find another location for the steam engine exhaust other than a pit like the one used in the old lab. According to Thomas Jefferson, the pit "was a messy, dank place that always smelled to high heaven." Gleason and Jefferson both advocated a condenser beside the small steam engine, and Jefferson verified that would work after seeing one at the Illinois Institute of Technology.[36]

A student from the early 1960s recalled that Gleason "was easily the senior member of the staff, but demonstrated a high degree of knowledge of industry. By contrast today, most of us young engineers were apprehensive of 'the first job task' in the real world. Could we really do what industry wanted? We survived, of course, but Jim Gleason provided the awareness and training that made the transition to industry far easier." A student in Gleason's machine design class remembered he would "announce that his test was closed book, hand out the test, and leave the room. He thought that the honor system was good. The trouble was that there were some Razorback footballers that thought otherwise. They took an open-book approach to a closed-book test. Gleason caught on and simply made the test open book, but changed the kinds of questions. The questions focused upon understanding the principles underlying the test problem, for without that knowledge, the student could not even figure out where to open the text. This experience led me to go for my Arkansas EIT exam armed with only my slide rule and my book of mathematical formulas. I smiled when one of the students pulled all of his texts in a Radio Flyer wagon over to his seat. I passed."[37]

Chapter 3

World War II and the Early Postwar Period, 1941–1958

The Paddock Years Continue

The 1940s were a time of extremes for the Mechanical Engineering Department as the war first took students away, then the GI Bill resulted in students overwhelming the staff. By the late 1940s, students who had graduated in May with a BSME would sometimes end up teaching right away to help out with the faculty shortage. By the early 1950s, enrollment numbers had leveled off, and the department continued to make strides in laboratories improvements as students continued the tradition of quality participation in ASME.

Department Head
Russell G. Paddock, 1941–1958

Faculty
1. Francis J. Daasch, BSME, Iowa State College; MSME, University of Minnesota; Associate Professor (1942–1944)
2. Hubert J. McAulay, BME, University of Detroit; Assistant Professor (1946–1947)
3. Cecil Oran Cogburn, BSME, University of Arkansas; Instructor (1947–1950), Assistant Professor (1950–1954), Associate Professor (1954–1964), Professor (1964–1984), University Professor (1984–1989)
4. William Foy Faires, Instructor (1946–1948)
5. James Otis Brown, BSME, University of Arkansas; Instructor (1946–1947)
6. James Ann Toone, BSME, University of Arkansas; Instructor (1947–1951)
7. Horace William Risteen, BSME, ME, University of Wisconsin; MS, Michigan College of Mining and Technology; Professor (1947–1951)

8. Robert Leroy Jeske, BS, MS, Oklahoma A&M College; Instructor (1947–1949), Assistant Professor (1949–1957), Associate Professor (1957–1990)
9. Grover W. Hughes, BSME, University of Oklahoma; Instructor (1947–1952), Assistant Professor (1952–1953)
10. Garland Samuels Jr., BSME, University of Arkansas; Instructor (1947–1948)
11. James E. Worsham, BSME, Vanderbilt University; Instructor (1948–1951)
12. Arthur V. Houghton, BS, University of Illinois; Instructor (1948–1950)
13. Reh L. Smith, BS, Oklahoma A&M College; Instructor (1949–1952)
14. James Jerome Kennedy, BSME, Tulane University; Instructor (1950–1951)
15. Haight W. Gurney, BS, University of Arkansas; Instructor (1952–1957)
16. Glover Edward Bagby, BSME, University of Arkansas; Instructor (1953–1954)
17. Marvin William Burnham, BSME, University of Kansas; Instructor (1954–1957)
18. Robert Blair Lucke, Instructor (1954–1957)
19. Frank K. Deaver, BSChE, MSME, University of Arkansas; PhD, University of Minnesota; Assistant Professor (1955–1960), Associate Professor (1960–1969), Professor and Department Head (1969–1980), Professor (1980–1984)
20. Lester R. Redmond, BSME, University of Arkansas; Assistant Professor (1956–1957)
21. Elmo M. Scott, BA, Hendrix College; Instructor (1957–1960)
22. Eugene Elton Billingslea, BS, United States Merchant Marine Academy; BSME, University of Arkansas; Instructor (1957–1961)
23. Lee L. Denny, BSIE; Instructor (1957–1958)
24. Omer L. Zillman, Instructor (1957–1973)

Many of the instructors hired during the late 1940s were recent graduates of the university. As enrollment increased with the influx of veterans, the department had to hire faculty as quickly as possible. By the

Students working on a P-38 fighter plane the university had acquired for aeronautical students. (Mullins Library Special Collections, photo collection #1835)

1951–1952 academic year, the faculty had decreased as enrollment stabilized.

In addition to hiring more faculty members, the college had made several departmental changes by the late 1940s. A new department, Engineering Mechanics, was organized for the 1948–1949 academic year to coordinate courses in mechanics that fell somewhere between physics, math, and professional and design courses of the several engineering departments. Courses included mechanics of materials, statics, dynamics, mechanics of fluids, and theory and laboratory. Similarities in course objectives found in several departments also led to the reorganization. Formerly part of the four existing engineering departments, Agricultural Engineering and Industrial Engineering became distinct departments beginning in the spring semester of 1949. The College of Engineering changed requirements regarding humanities and social studies credits by allowing only twelve ROTC credits to be completed in addition to humanities credits instead of allowing for substitution of ROTC credits in their place.[1]

The Aeronautical Department, headed by Associate Professor Gleason, also appeared around the same time. The department had nearly the same requirements as the mechanical engineering degree, which had not changed considerably since before the United States entered the war. Requirements for seniors who enrolled in the Aeronautical Department differed the most from those for the BSME.

Aeronautical Engineering Curriculum, 1946

Freshman
Fall, 17 credit hours
Engl, Composition, 3 credits
Chem, General Chemistry, 4 credits
Math, College Algebra, and
Math, Trigonometry, 6 credits
Drawing, Mechanical Drawing, 2 credits
ME, Shop, 1 credit
ROTC, Military Art, 1 credit

Spring, 18 credit hours
Engl, Composition, 3 credits
Chem, General Chemistry, 4 credits
Math, Analytics, 5 credits
Drawing, Descriptive Geometry, 2 credits
ME, Shop, 1 credit
GE, Engineering Problems, 2 credits
ROTC, Military Art, 1 credit[2]

Sophomore
Fall, 18 credit hours
Phys, General Physics, 5 credits
Math, Calculus, 4 credits
Drawing, Mechanical Drawing, 2 credits
Econ, Principles of Economics, 3 credits
Engl, Composition, 2 credits
ME, Pattern Shop, 1 credit
ROTC, Military Art, 1 credit

Spring, 18 credit hours
Phys, General Physics, 5 credits
Math, Calculus, 4 credits
Drawing, Mechanical Drawing, 2 credits
Econ, Principles of Economics, 3 credits
Engl, Public Speaking, 2 credits
ME, Machine Shop, 1 credit
ROTC, Military Art, 1 credit[3]

Junior
Fall, 18 credit hours
ME, Mechanics, 5 credits
ME, Thermodynamics, 3 credits
ME, Mechanism, 3 credits
ME, Seminar, 1 credit
EE, Electrical Engineering, 3 credits
Aero, Aeronautical Engineering, 3 credits

Spring, 19 credit hours
CE, Materials, 5 credits
ME, Thermodynamics, 3 credits
ME, Kinematics, 2 credits
EE, Electrical Engineering, 3 credits
EE, Electrical Engineering Lab, 2 credits
Aero, Aeronautical Engineering, 3 credits
Aero, Aeronautical Engineering Lab, 1 credit

Senior
Fall, 17 credit hours
CE, Hydraulics, 4 credits
CE, Airplane Structures, 3 credits
Aero, Aeronautical Design, 3 credits
Aero, Aeronautical Engineering Lab, 2 credits
ME, Mechanical Engineering Lab, 2 credits
GE, Law and Contracts, 3 credits

Spring, 18 credit hours
CE, Materials Lab, 1 credit
CE, Airplane Structures, 3 credits
ME, Metals, 3 credits
Aero, Aircraft Engines, 3 credits
Aero, Aeronautical Design, 2 credits
ME, Industrial Engineering, 3 credits
ME, Seminar, 1 credit
ME, Mechanical Engineering Lab, 2 credits[4]

By the late 1940s, some mechanical engineering students thought changes should occur in the curriculum. An anonymous open forum in the *Arkansas Engineer* shared the concerns from each department. A senior mechanical engineering student expressed fellow students' displeasure on three topics. Students thought that the shops focused too much on basic work and not enough on specific operations they might be expected to know for a job. Some thought that inspection trips ought to be included in the degree requirements again. Also, the college as a whole needed to require more courses from the Colleges of Business and Arts. By the 1948–1949 academic year, more humanities electives had been added to the curriculum. The college reduced strength of materials to four hours and the course title changed to Mechanics of Materials, which also changed from a one- to a two-semester course.[5]

An open forum in the *Arkansas Engineer* in the spring of 1951 included graduating seniors from each department. The consensus among mechanical engineering seniors was that too much emphasis had been placed on steam power that resulted in a lack of course options by their senior year. They advocated allowing seniors to choose other courses of similar caliber as steam power such as refrigeration and air conditioning. Also, they thought the lab workload was too much for the amount of credit they had received. They also expressed displeasure with the rigid curriculum that forced students to make up specific classes that sometimes extended their degree for another year. Some students simply chose another major rather than complete the mechanical engineering degree because of the expense of enrolling for another year. They also thought that professors did not grade assignments and tests fast enough to let students know their standing.[6]

A few years before, a similar concern about the length time to complete a BSME had been expressed in 1948. In an article in the *Arkansas Engineer,* a junior mechanical engineering student advocated the consideration of a five-year plan for engineering students similar to those at other universities. According to the student, this would help with the massive amount of work the typical upperclassman encountered.[7]

By the end of the Paddock era in 1958, the Mechanical Engineering Department continued to offer seniors the option of enrolling in aeronautical courses their final year, but the separate Aeronautical Department no longer existed. The college had also begun an arrangement with the Fayetteville Flying Service to allow students to receive flight training. Another new adjustment allowed students to specialize in air science while in the ROTC program. After taking freshman and sophomore air science, a student could enroll in advanced air science training during the junior and senior year with a four- to six-week training camp over the summer. After graduation, the student could become a commissioned officer in the United States Air Force Reserve.[8]

The separate engineering departments now had control once again over all four years of undergraduate coursework. Many of the departments required the same classes, including freshman orientation that familiarized students with the purpose of each college so they could decide if a different department might be better suited for their goals. A new requirement for technical electives appeared in the senior year. The elective, under approval from the head of the department, could be fulfilled by enrolling in an upper-level course in the college or in science departments such as mathematics and physics.[9]

Curriculum, 1957

Freshman

Fall, 17 credit hours

Engl, Composition, 3 credits
Chem, General Chemistry, 4 credits
Math, College Algebra, 3 credits
Math, Plane Trigonometry, 3 credits
IE, Engineering Drawing I, 2 credits
ME, Elementary Pattern Shop and Foundry, 1 credit
ROTC, Military or Air Science, 1 credit

Spring, 18 credit hours
Engl, Composition, 3 credits
Chem, General Chemistry, 4 credits
Math, Analytic Geometry, 3 credits
Math, Calculus I, 3 credits
ME, Elementary Forge and Machine Shop, 1 credit
IE, Engineering Problems, 2 credits
GE, Freshman Orientation, 1 credit
ROTC, Military or Air Science, 1 credit

Sophomore
Fall, 18 credit hours
Phys, Engineering Physics, 3 credits
Phys, Physics Lab, 1 credit
Math, Calculus II, 3 credits
Econ, Principles of Economics, 3 credits
IE, Descriptive Geometry, 2 credits
ME, Pattern Shop, 2 credits
A&S, Humanities, 3 credits
ROTC, Military or Air Science, 1 credit

Spring, 18 credit hours
Phys, Engineering Physics, 3 credits
Phys, Physics Lab, 1 credit
Math, Calculus III, 3 credits
Econ, Principles of Economics, 3 credits
EM, Statics, 3 credits
IE, Engineering Drawing II, 2 credits
ME, Machine Shop, 2 credits
ROTC, Military or Air Science, 1 credit

Junior
Fall, 17 credit hours
EM, Dynamics, 3 credits
ME, Thermodynamics, 3 credits
ME, Mechanism, 2 credits
ME, Mechanical Lab, 1 credit

Spch, Principles of Effective Speaking, 2 credits
ME, Seminar, no credit
EE, Electrical Circuits and Machines, 3 credits
Humanities, 3 credits

Spring, 17 credit hours
EM, Mechanics of Materials, 3 credits
ME, Thermodynamics II, 3 credits
ME, Kinematics, 3 credits
ME, Mechanical Lab, 1 credit
ME, Seminar, 1 credit
Phys, Elements of Atomic Physics, 3 credits
EE, Electronics, 2 credits
EE, Electrical Equipment Lab, 1 credit

Senior
Fall, 19 credit hours
ME, Steam Power Plants, 4 credits
ME, Machine Design I, 3 credits
ME, Advanced Mechanical Lab, 2 credits
EM, Mechanics of Fluids, 3 credits
ME, Seminar, no credits
EM, Mechanics of Fluids Lab, 1 credit
Technical Elective, 3 credits
Humanities Elective, 3 credits

Spring, 18 credit hours
ME, Internal Combustion Engines, 4 credits
ME, Machine Design II, 2 credits
ME, Advanced Mechanical Lab, 2 credits
ME, Metals, 3 credits
ME, Seminar, 1 credit
Technical Elective, 3 credits
Humanities Elective, 3 credits[10]

By the early 1940s, hostilities in Europe and in the Pacific had resulted in young men being called up for the draft even before the attack on Pearl

Harbor. Although students could seek exemption, grades had to be kept at a certain level. On the issue of draft deferment for engineering students, Dean George Stocker noted that poor grades would result in a report made to the local draft board. A year later, after the United States had entered the war, Dean Stocker found himself in a different situation. Instead of students attempting to barely stay in college and avoid the draft, some began to leave in order to receive a better position after the draft age had been lowered. Dean Stocker advised students might do better than they thought in school and any amount of education would help with military placement. By 1944, he expressed concern about the rapid pace at which students were expected to graduate considering the difficulty of quickly mastering technical material like engineering.[11]

A student who began the BSME program in 1942 recalled that he wanted to wait until after he graduated to join the military. Education waivers for engineering students made this possible, but by 1944, they were no longer available. He volunteered for the navy rather than being drafted and taught radar and sonar technicians at the Naval Research Lab during 1946.[12]

University enrollment increased during the early years of World War II as almost 2,500 students entered programs such as the air corps and specialized training units. By the 1944–1945 academic year, that number had dropped to 366 as the students completed the programs, and women made up almost all of the 2,068 civilian students. This fluctuation in enrollment overwhelmed the faculty members at first since many had entered the service, but as the military enrollment decreased, upper-level civilian students found themselves with plenty of faculty for instruction.[13]

The number of students in classes decreased also as a result of poor attendance by civilians. A former mechanical engineering student recalled that one professor would call to check if enough students would be attending his 8:00 A.M. class. The student's roommate at the Theta Tau house missed class so much that he nearly flunked. In order to get to class on time, he devised a system that resulted in an electric shock after the tenth ring of his alarm clock. Another former mechanical engineering student recalled that class sizes of fifteen to twenty students in September of 1942 had decreased to four or five by the end of October. He was the only student still enrolled in machine shop class by Thanksgiving.[14]

As the university made substantial schedule changes to accommodate those entering the military, the administration also implemented programs for those remaining on the home front. Before the United States actually entered the war, the university, along with the federal government, established the Engineering, Science, Management, Defense Training program to train men and women to work in defense industries. Once the United States actually entered the war, the name of the program changed to Engineering, Science, Management, War Training (ESMWT), but the program still trained men and women for the same purpose. Although instruction occurred on the campus of the university, approximately twenty-five hundred workers in the towns of defense industries usually received a considerable amount of the courses taught in the program. Extension and correspondence courses grew in general as servicemen and students in towns that had been depleted of teachers took advantage of long-distance education offerings.[15]

To accommodate those students enlisting in the military after graduation, the university instated a quarter system to allow for faster matriculation. New students could be accepted four times during the year, in June,

Camp Neil Martin under construction near present-day Hotz Hall, ca. 1943. (Mullins Library Special Collections, photo collection #1248)

September, January, and March, and a student could complete a degree in three years instead of four. The academic year expanded the amount of summer courses offered to more than 350. To accommodate the larger student population, the university constructed Camp Neil Martin and turned two dormitories, Razorback Hall and Mary Anne Davis Hall, into barracks for those involved with military training.

Throughout the war and afterward, the university constructed many other barracks and housing units used first by recruits, then veterans and, if they had them, their families. The enrollment continued to grow rapidly at the University of Arkansas as it did at campuses all across the country in post–World War II America, a problem university officials welcomed. As a result, the temporary housing west and northwest of the campus became permanent, and some of the structures remained standing and in use through the late 1960s and 1970s.[16]

The College of Engineering had 572 students enroll for the 1945–1946 academic year. That number nearly doubled the following year as 1,029 students sought an engineering degree. After several years out of the classroom, many veterans required some time to remember what they had learned before leaving for war. A mechanical engineering student during that time recalled that he had taken differential calculus in 1942 and had to leave shortly after finishing that class. Returning in 1946 for his junior year, he had to take integral calculus, thermodynamics, and mechanics. Thermodynamics and mechanics both required knowledge of integral calculus, and that required a thorough knowledge of differential calculus and trigonometry, which he had taken back in 1942. Fortunately, Professor Paddock understood the gap that had developed and worked with the students to ensure success.[17]

Some students chose to look off-campus for housing. Since some returning veterans had a family, housing on campus did not suit those needs immediately after the war. One mechanical engineering student lived in a small summer camp building on Mount Sequoya with his family. However, they soon found on-campus housing when the university converted army barracks into apartments, known as Lloyd Hall, near the football stadium. Another student seeking a BSME degree lived with his wife in the trailers near present-day Hotz Hall for two years until they moved to a basement apartment on Mount Sequoya. After living in a trailer susceptible to leaks during rainstorms and frozen pipes in the winter, he and his wife welcomed the opportunity for improved living arrangements.[18]

Camp Leroy Pond was located near present-day Bud Walton Arena, ca. 1945. (Mullins Library Special Collections, photo collection #1205)

Dean George Branigan, Dean Stocker's successor since the fall of 1948, encountered enrollment problems he attributed to postwar military activity. He expressed concerns over the drop in enrollment of engineering students and attributed the decline partly to Selective Service withdrawing engineering students. An article appeared in *Fortune* that discussed the shortage of engineers in the workforce. As Dean Branigan had suggested, the article speculated that draft policies might have resulted in a shortage of engineers to fill a number of vacant positions. The Soviet Union reportedly did not have the same vacancy problem in the engineering field as the United States did.[19]

The public concern over low engineering enrollment numbers appeared to have some effect, as Dean Branigan reported a 29 percent increase in freshman enrollment in engineering colleges nationwide based on numbers from the fall of 1952 compared with those from the year before. He mentioned that the University of Arkansas did much better, posting a 90 percent increase. Of the freshman class, 4.6 percent enrolled in the College of Engineering in 1952 while only 2.8 percent did so before

World War II. By the fall of 1955, the college reported a record enrollment of 832 students. Mechanical engineering students comprised 202 of those students. Electrical engineering had the most students, 264.[20]

By the middle of the 1950s, veterans from the Korean War would make up a significant amount of the student body. Usually, these students were older than the more traditional college population. One mechanical engineering student recalled that the veterans "thought it was important to study and meet the curriculum as laid out in the catalog—144 hours in four years, which averaged eighteen hours per semester. I had $800 per school year, and that was it. So, I emulated the vets. Considering that a student spent at least one hour of study for each hour of class, that was at least a thirty-six-hour week. So for us, there was not much activity other than studying."[21]

After the United States entered into World War II, the ASME continued to plan and hold meetings and smokers. ASME district representative Professor Howard E. Degler from the University of Texas had visited in the fall of 1941 and noted that student sections did not spend all of the money provided by the national office. The officers pledged to hold more smokers with hard cider and invite men in the engineering field to speak at these events. To this effect, the student section held a smoker in January of 1942 at the ECHO house where members listened to three presentations. Before Professor Price left for Michigan in 1942, the ASME section held a smoker in his honor for serving as the faculty sponsor. According to the Razorback at the time, smokers were held "so that the boys can become better acquainted by exchanging the latest in engineering developments and varicolored jokes."[22]

After the war, ASME continued with regular meetings and presentations, and more papers began to discuss the idea of space travel. Bill Passarelli predicted travel to the moon by rocket would be possible in ten years. Another talk discussed the German V-2 rocket and possible developments of a similar project in the United States. By 1947, membership had doubled from the previous year to 84. The number continued to rise, and ASME claimed 150 members by the fall of 1948. The large membership led to the section separating into two subsections. The seniors gave seminar talks the first semester, followed by the juniors in the spring semester.[23]

During the spring of 1953, the section traveled to Lincoln, Nebraska,

ASME members. (*Razorback*, 1956)

to participate in the annual ASME conference and won several prizes. Albert S. McDaniel won a second prize of $25 for his paper on "Shell Moldings." The section received a plaque for having the largest percentage of eligible members enrolled in the section and a trophy for traveling the furthest distance to the conference. The section split a $40 prize with Kansas State for the largest percentage of enrolled members participating in the conference. The participants also made inspection trips to Cushman Motor Scooter Company and the Nebraska Tractor Testing Laboratories.[24]

During the 1949–1950 academic year, construction began on an addition to Engineering Hall to handle the increased enrollment in the university and the College of Engineering. By the fall of 1950, the construction was complete, and the engineering departments became more centrally located when the chemical engineering department moved into the building. The $125,000 addition also doubled the physical size of the engineering library, allowing the collection to increase from 17,000 to 37,000 volumes.[25]

As students returned from the Korean War, they found the mechanical engineering lab with plenty of equipment. The lab had three steam engines that included a Ridgeway simple steam engine, a McEwan

Engineering Hall addition under construction, 1950. (Mullins Library Special Collections, photo collection #1055)

tandem compound, and a Fitchburg uniflow. The Waukeshaw bus engine and a General Motors engine were used with a dynamometer. The Foos gasoline engine allowed for multiple tests, such as water injection. The single-cylinder gasoline engine Elsie had adjustable timing for both exhaust and intake valves. A new diesel generator engine was set up to use seawater as the coolant. The Buda four-cylinder, four-cycle diesel engine featured a water brake. The Fairbanks Morse diesel simulated the load of a commercial ice plant through its connection to a two-stage air compressor. A former mechanical engineering student recalled that when the diesel was fired up for experiments, it shook the entire building. Apparently, the noise sounded all too familiar to a professor from California who instructed his class to leave when he heard the engine for the first time in the belief it was an earthquake. Besides the sometimes unbearable noise, working with equipment did not always have the desired effect as one professor recalled, "There was a lot of trouble running labs and getting a clean, clear-cut impression of what you were trying to prove." As a result of the tedious work that labs often required, the textbook commonly used during that time by authors Charles F. Shoop and George L. Tuve was nicknamed "Stoop and Shove" by students.[26]

In an open forum in the *Arkansas Engineer* on lab courses in each of the departments, one student thought that the mechanical engineering labs did not utilize the equipment enough, and the lack of labels on

Student adjusting lab equipment. (Mullins Library Special Collections, photo collection #1839)

equipment parts, such as steam pipes, prevented students from the full benefits of experiments. The sector of the economy that developed due to the growing automobile industry made valuable experience with the engine equipment that much more important. By the fall of 1950, the Automobile Manufacturers' Association estimated the auto and auto parts sector created $25 million annually, an encouraging amount for mechanical engineering students. DuPont actively recruited mechanical engineering students by advertising in the pages of the *Arkansas Engineer,* pointing out problems they required help solving and the facilities available in which to carry out the work. The company needed high-speed silting equipment, equipment to operate at pressures up to 4500 psi, and large air-conditioning systems for the manufacture of certain products. Dupont spent $3.5 million on its Wilmington shops, which covered 300,000 square feet and included a foundry and a pattern shop. Over eight hundred employees worked in the shops, and potential output was estimated at $6 million annually.[27]

New and old problems involving Engine Week confronted administrators and students. The pranks between the engineering and agriculture colleges that had occurred for many years on the respective festival days culminated in significant property damage in the spring of 1947. According to a former mechanical engineering student, "there may have been some eggs thrown at the aggie house, Alpha Gamma Rho, and some green paint accidentally spilled. Retaliation came with an invasion of the Theta Tau house as an engineering student was overwhelmed and given a free haircut." This resulted in Dean Stocker requesting students to refrain from future agitation by avoiding excessive celebrations that might lead to such confrontations with the agriculture students. The competition for St. Patricia reached a level that some considered unacceptable by the 1950s. An editorial in the *Arkansas Engineer* admonished the engineering students for encouraging the female contestants to exhibit themselves in a way that bordered "on the burlesque."[28]

A mechanical engineering student in the late 1940s recalled how ASME continued to impress the open-house crowds on Engineer's Day: "The ASME members contacted the New Departure Bearing Division of General Motors and requested the use of their ball sphericity demonstrator during our Engineer's Day activities. GM agreed, and the display was the star of our mechanics laboratory. The mechanism was in a beautiful oak cabinet, which had two openings near its top and two angled polished steel anvils facing towards the center where a single anvil had an angular face facing the cabinet. Once connected to a power outlet, a stream of half-inch balls started dropping out of the two upper spouts, bouncing on the two anvils, bouncing to the center anvil and finally bouncing into a center hole. The continuous stream of perfect balls demonstrated precision manufacture. If a defective ball were placed in the hopper, the imperfect ball would wildly bounce out of the beautiful, dynamic symmetry and not return to the hopper. People would watch this display for long periods hoping for a mistake, but the only way to cause a fault was to put in a discrepant ball."[29]

The continued growth of the United States military after World War II benefited the university and the Mechanical Engineering Department with the 1945 establishment of the Ordnance-Arkansas (ORDARK) research department, which received funding through one of many research and development contracts made available to universities nationwide. Although the scientific research conducted in the department typically had military applications, a second mission focused on innovation for the industrial sector. The contract included funding for graduate fellowships, and a new building on Dickson Street that opened in 1947 with modern equipment that could be used by other related university departments. The department also provided employment opportunities for graduates in mechanical engineering in several positions. Originally, the work done within the ORDARK department had to remain classified, but this did not please many involved with the university since students could not benefit from the research. In order to allow student involvement with the research, the university established the Institute of Science and Technology.[30]

The goals of the Institute of Science and Technology included not only providing more research opportunities for students and science-related departments but also information for state agencies involved with economic development. Not all university departments believed the institute addressed their concerns with ORDARK, but research did increase on campus. The increased research also helped make the case for improvement of buildings and facilities, and university president Lewis Webster Jones believed the institute organized research so that duplication of the same research did not occur in different departments.[31]

As well as providing a rationale to development graduate programs, the industrial development in Arkansas provided many opportunities for ASME field trips. In March of 1953, ASME members visited the Arkansas Power and Light plant in Forrest City, the steam plant at Lake Catherine, the hydro plant at Lake Hamilton, and the Reynolds Metal Aluminum Plant between Hot Springs and Malvern. The university began to make plans by the late 1940s to implement a doctoral program due to the increasing requirement of such a degree in many industries in Arkansas. The university had hesitated in the past to proceed with plans to expand graduate studies until proper research facilities for the sciences could be constructed.[32]

The Postwar University

The University of Arkansas had recognized the need for more graduate education programs beginning in the late 1940s. Since the university did not offer PhD or EdD programs, students in Arkansas either did not receive the degree or left the state for another institution, most often never returning to the state. As a result, certain areas of the state suffered from the lack of a highly educated workforce. Although the university offered coursework leading to the MA degree, even those requirements had to be restructured to be on a similar level as other institutions before any more programs could be added. By the fall of 1950, the university had inaugurated its PhD and EdD programs offering candidacy in Education, English, Philosophy, History, Economics, Chemistry, and Biochemistry.[33]

At the time of World War II, Fayetteville had a population of 8,200, and many felt the effects of the conflict as fathers and sons left for war or families moved to states such as California where large defense industries had positions to fill. Fayetteville industries such as canning and lumber also stepped up production to meet the demands of a wartime economy. In the late 1940s, Fayetteville's downtown reflected the postwar production of manufactured goods as stores that lined the streets filled their display windows. However, the renewed interest in shopping along with an influx of new residents and their private automobiles began to cause parking problems downtown by the beginning of the 1950s. The commercial district began to extend north on College Avenue as a result of the economic boom and the congestion problems that accompanied it. Students looking for distractions could head to the Tee Table, south of Fayetteville

College Avenue in Fayetteville, ca. 1940. (Mullins Library Special Collections, Kent R. Brown Collection, Malone #4)

on US 71, as well as Schuyler Town, as lower Dickson Street was called. For a more formal setting, one could go to Tontitown or the A.Q. Chicken House in Springdale.[34]

By the late 1950s, the university had serious problems of its own regarding the private automobile. With about one car registered for every two students, parking regulations became a new facet of university administration. Such things as planning for new lots, lot designations, and student infractions related to automobile policies had to be considered. As in other college towns across the country, the University of Arkansas campus began the shift from one dominated by walking to one yielding to the automobile.[35]

The strong postwar economy in the United States presented other problems for university officials as high-paying industrial careers attracted many highly qualified faculty members, and the Mechanical Engineering Department, along with the other departments in the college, had to make adjustments. According to the North Central Association, which accredited the university for several doctoral programs, the salaries the university offered experienced faculty could be blamed for a lot of the retention problems. The association warned the loss of such faculty members would greatly hinder the research possibilities for graduate students. Dean Branigan recognized the teaching shortage in engineering colleges and advocated attempts to close the pay gap between the industrial and academic job markets.[36]

However, funds for research and faculty salaries would become more available as the United States began to respond to the threat posed by the military and science programs in the Soviet Union. The successful launch and orbit of Sputnik in October of 1957 set the wheels in motion to promote the rapid advance of technology in the United States. A concerted effort from both private and public sources to release funds to provide the necessary equipment, manpower, and facilities resulted, and universities suddenly had the financial wherewithal to compete with private industry for engineers and scientists. By 1968, universities' expenditures nationwide on research and development totaled $2.6 billion compared with $334 million in 1953, and universities awarded more graduate degrees during the same period, many of which went to students who found plenty of employment opportunities in the federal government.[37]

Changes on the campus of the University of Arkansas in the post–World War II period involved more than just academics. In February of 1948, Silas Hunt enrolled in the University of Arkansas School of Law to become the first person of color to do so. The university had attempted to instruct black students when the school opened in 1872 based on a resolution by the board of trustees, but,

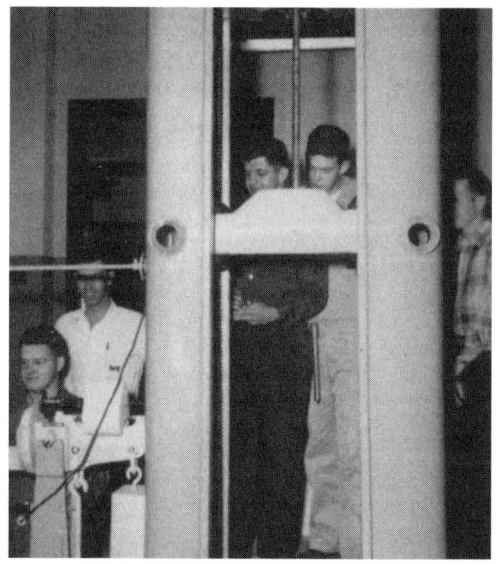

Students in the Mechanical Engineering Lab. (*Razorback*, 1956)

apparently, no blacks actually obtained degrees. After the Branch Normal College opened in Pine Bluff in 1875, pressure for black admittance to the university dissipated. By the late 1940s, legal proceedings had begun in Oklahoma and Texas to force both state universities to admit black students. Dean of the Law School, Robert A. Leflar, viewed the integration of law schools inevitable based on the situation in the neighboring states and advocated segregated instruction for black students at the university. Also, the 1872 resolution by the board of trustees had never been altered, so no barrier existed in the language of the university's policies to prevent entrance of black students. Hunt registered and began studies at the university but had become quite ill by the summer and died from tuberculosis at the Veterans' Hospital in Springfield, Missouri. Hunt opened the door, however, and Jackie L. Shropshire registered in the fall of 1948 and was followed by three other students the next year. During these years, the campus gradually accepted the new students evidenced by the elimination of many of the restrictions put on Hunt when he had enrolled at the university. Although claims of mistreatment arose many years later concerning the integration of the Law School, the situation remained unusually calm for the period, and some white students actively interacted with the black students, even attending the segregated classes with Hunt.[38]

The College of Engineering regained an important research facility when the Engineering Experiment Station resumed plans for full operation during 1949. As a result of the great amount of resources and manpower contributed to World War II, the station limited the research that had been ongoing since the early 1920s. The board of trustees formerly established the station on November 6, 1920, to be managed by the president of the university, the dean of the College of Engineering, and the engineering department heads. After the station commenced operations, many of the projects focused on the best utilization of the state's resources primarily for economic development, the results of which the station published in a bulletin. Once the university organized the campus to handle the large influx of students after the war, attention could be focused on resuming the mission of the station. The Engineering Experiment Station also fulfilled a need at the university to contribute to the expansion of industry. Industry no longer relied only on organization to advance new ideas, and in post–World War II America, many in both the private and public sectors realized college campuses could be the main source of those ideas.[39]

The Institute of Science and Technology, organized in 1948, began to coordinate this interest in university research and, by 1952, had a budget of $532,000, most of which came from federal government contracts. The institute also determined the kind of research that would take place in the humanities departments as well, which caused concern, as some department heads believed the institute had too much control over university research. After a reevaluation in 1953 determined that the educational mission of the university may have been supplanted too much by research, the research projects and specialists within the institute were absorbed into the appropriate departments, and the institute ceased to exist by June of 1955. However, the institute did prove that the university required a single office to handle research inquiries and the financial resources they would bring. Provost Lewis H. Rohrbaugh decided to turn this task over to the dean of the Graduate School, V. W. Adkisson, and named him research coordinator. The Office of the Research Coordinator, a part of the Graduate School, assisted a great deal in the success of the university to handle the increasingly large research grants during the 1960s. By 1970, the university spent $15,456,794 on research projects that included almost every department.[40]

Students working with equipment in the Mechanical Engineering Lab. (*Razorback*, 1957)

Despite the increased attention toward the funding of science and engineering after World War II, the university continued to support the arts. In 1950, the first performance opened in the new $1,030,922 fine arts center complex designed by Arkansas architect Edward Durell Stone. The buildings included a theater, a concert hall, an art gallery, and classroom space connected together. The formal dedication on May 5, 1951, was part of a week-long celebration of the arts that included concerts and exhibits.[41]

Athletics

The university decided to take steps to modernize the athletic program during the war years when an athletic committee, mostly composed of alumni, developed in 1943. The board of trustees also became involved when the chairman selected a group of Arkansas citizens to serve on an athletic director search committee, and they hired John Barnhill in December of 1945. Barnhill also served as head football coach after three different coaches held that position from 1942 to 1945. He quickly found success after leading the Razorbacks to two consecutive bowl appearances in the Cotton Bowl in 1947 and the Dixie Bowl Inaugural in 1948.[42]

After two consecutive 5-5 seasons in 1948 and 1949, Barnhill decided to hand the head football coaching duties to someone else and concentrate on the administration of the athletic department. The next three coaches all had varying degrees of success. Otis Douglas coached the team from 1950 to 1952 and never established a winning program. Bowden Wyatt followed Douglas and achieved success in his second season when the Razorbacks won the SWC championship and played in the Cotton Bowl. Wyatt departed in January of 1955 to accept the head coach position at Tennessee, and Jack Mitchell replaced him. Mitchell coached the team for three years and produced winning squads, but decided to leave Arkansas for another position. Mitchell's departure opened the door for Frank Broyles to accept the head coaching position, and after one disappointing season, Arkansas finished in a three-way tie for first place in the SWC. Broyles would remain for many years as the head coach and lead the team to national prominence.[43]

The basketball team did not experience the same level of success as the football program after World War II. Glen Rose assumed head coaching duties once again in the 1950s and guided one team to the SWC championship in the 1957–1958 season and had a few second-place teams. However, the successful seasons did not occur as often as they did in football, which had begun to overshadow basketball by the 1960s as the most popular university sport.[44]

The Razorback track team began competing in the late 1890s and held the SWC track meet at the university in the spring of 1922. However, the team did not finish well in that meet and found little success until the 1950s when the cross-country teams began winning championships on a regular basis. Other athletic teams have existed and found some success, while others have been discontinued or remained out of intercollegiate competition. A wrestling team competed well at Arkansas for many years until the SWC dropped the sport in 1925. By the 1960s, the golf team had improved and competed for conference titles, and swimming, which began in 1966, quickly gained attention as teams competed well in the conference. Tennis had not become a highly competitive sport by the 1960s, and sports such as boxing and fencing remained club activities.[45]

For mechanical engineering students, studies could often restrict extracurricular activities, or in some cases, a family was sometimes a concern as well. However, the challenging employment opportunities that resulted from the country's desire to stay ahead of the Soviet Union in the cold war usually paid off for those who studied hard. Also, the development of more graduate programs and the funding that accompanied those provided more opportunities for high scholarship than ever before. The opportunities to improve mechanical engineering and other engineering programs would become more numerous in the 1960s as even more money became available for research and facilities.

Biographical Sketches

Cecil Oran Cogburn, BSME (1942), MSME (1942), University of Arkansas; DIC (1964), University of London; PhD (1970), University of London; Instructor (1947–1950), Assistant Professor (1950–1954), Associate Professor (1954–1964), Professor (1964–1984), University Professor (1984–1989)

Cecil O. Cogburn taught Mechanisms, Kinematics, and Nuclear Heat Transfer, in addition to brief stints in Pattern, Wood, and Machine Shops. Two professors died suddenly and unexpectedly in the late 1940s, and Cogburn was asked to teach shop courses, which caused considerable consternation for the young instructor as the department and the college were already overwhelmed by the influx of students on the GI Bill. According to Cogburn, "We had students that, boy, I will tell you, they were eager. Saturday classes, nights, nothing stopped those young fellows."[46]

The second time he studied in England, Cogburn received a fifteen-month National Science Foundation faculty fellowship. He returned to the university after being on leave for two years and completing his doctoral work in London. As a result of his work in London in the field of

Cecil O. Cogburn, ca. 1975. (Mechanical Engineering collection)

nuclear research during his first visit to Britain, Cogburn taught the first nuclear engineering courses in the mid-1960s. In addition to teaching Basic Nuclear Engineering, Cogburn recalled setting up the first lab that had "a little bit of equipment to study radiation effects, distance effects in shielding, and metal, wood, and concrete effects in shielding. I as remember, we had one little recording device—a multichannel analyzer it was called."[47]

Several students recalled Cogburn as a demanding instructor, but many students appreciated that style. One student from the early 1960s thought that although Cogburn's grading system was very "tough in the students' eyes—a 'C' in his class was easily an 'A' anywhere else—we have to give him credit because we did learn." Another student from a few years earlier remembered when Cogburn thought he was about to receive a chicken trophy at the Engineers' Banquet. A skit portrayed Cogburn's penchant for neatness on assignments, but the students awarded him the Outstanding Instructor Award instead, showing their appreciation for his style. Another student who took Machine Design I & II from Cogburn remembered he "was a demanding instructor, but I learned a lot in that class that helped my career."[48]

Robert Leroy Jeske, *BS (1947), MS (1947), Oklahoma A&M College;
Instructor (1947–1949), Assistant Professor (1949–1957),
Associate Professor (1957–1990)*

Courses that Professor Robert Jeske taught included Pattern Shop and Foundry. One student recalled making an "aluminum cast Razorback, which I still have in my office, and other neat items such as a nut cracker and desk nameplate." A student in the early 1960s remembered taking machine shop and foundry in some old Quonset huts. According to the student, Jeske "was pragmatic and encouraging."[49]

During the spring of 1980, Jeske and his wife visited automobile plants in the United States, Europe, and Japan. The purpose was to compare foreign assembly plants to those in America and attempt to determine if the

Robert L. Jeske, ca. 1981. (Mechanical Engineering collection)

plants' efficiency led to the growing number of imports in the U.S. automobile market. Jeske paid for the trip himself. Although classes had to be taken up by other professors, Ken Deaver, the department head at the time, supported his trip because Jeske had often done the same for other professors in the past.[50]

Chapter 4

The Jefferson Years, 1958–1969

The turmoil on the campus of the University of Arkansas as well as nationwide that developed in the 1960s had little impact on the Mechanical Engineering Department. The period of time that Professor Thomas B. Jefferson served as department head would undoubtedly be remembered most for the Mechanical Engineering Department's move into its own building. Mechanical engineering was not the only department in the engineering college or in the university to benefit from new facilities. New academic buildings and residence halls were built, dramatically altering the configuration of the campus. With the new facilities, the university hoped to attract better faculty and students and the financial resources for research that often accompanied them.

Department Head
Thomas B. Jefferson, 1958–1969

Faculty
1. Thomas B. Jefferson, BSME, Kansas State University; MSME, University of Nebraska; PhD, Purdue University; Professor and Department Head (1958–1969)
2. Vernon E. McBryde, BSIE, University of Arkansas; Instructor (1958–1959)
3. David O. Watts, BSME, Alabama Polytechnic Institute; Instructor (1958–1960)
4. Norman Cail, AB, BS, University of Arkansas; Instructor (1959–1962), Assistant Professor (1962–1963)
5. Dale K. Canfield, BSME, University of Arkansas; MSME, Georgia Institute of Technology; Assistant Professor (1960–1963)
6. Henry H. Hicks Jr., BSChE, University of Arkansas; MSE, PhD, University of Michigan; Professor (1961–1991)
7. Helmut Wolf, BSME, Case Institute of Technology; MSME,

PhD, Purdue University; Associate Professor (1961–1966),
Professor (1966–1972), Distinguished Professor (1972–1988)

8. Robert Lee Lott Jr., BSME; Instructor (1961–1962)
9. Jim H. Akin, BSME, MSME, PhD, University of Texas;
    Assistant Professor (1963–1968), Associate Professor
    (1968–1974), Professor (1980–1998)[1]
10. Frederick J. Crossett, BSEE, Drexel Institute of Technology,
    Pennsylvania; MS (Purdue University); Assistant Professor
    (1963–1964)
11. Arlis Jay McNemar, BSME, University of Oklahoma; Instructor
    (1963–1964)
12. David Milton Scruggs, BSME, MSEMtl, PhD, University of
    Michigan; Assistant Professor (1966–1969), Associate Professor
    (1969–1974), Professor (1974–1976)

Russell Paddock became emeritus professor after the 1957–1958
school year and was followed by Thomas B. Jefferson. A solid core of professors would improve in the 1960s with the addition of Professors Wolf,
Hicks, and Akin. By the end of the 1950s, the Mechanical Engineering
Department had 246 students, second behind electrical engineering with
416. These numbers corresponded to national trends that showed electrical, mechanical, and then civil engineering according to enrollment
numbers.[2]

The mechanical engineering curriculum in the late 1950s generally
followed two paths. One covered the science and art of machine design,
and the other focused on the science of heat power since most machines
operated, directly or indirectly, by some kind of heat engine. If a student
completed work in those fields, the department anticipated success for the
graduate operating the machinery in a specialized field such as aeronautics or air conditioning.[3]

The course requirements for freshmen and sophomores remained the
same for Jefferson's first year, but the department implemented changes in
the course requirements for juniors and seniors. Specifically, the junior year
included an inspection trip course for no credit, and credit hours shifted
so that the junior year required more than the senior year.[4]

Curriculum, 1959

Freshman
Fall, 17 credit hours
Engl, Composition, 3 credits
Chem, General Chemistry, 4 credits
Math, College Algebra, 3 credits
Math, Plane Trigonometry, 3 credits
IE, Engineering Drawing I, 2 credits
ME, Elementary Pattern Shop and Foundry, 1 credit
ROTC, Military or Air Science, 1 credit

Spring, 18 credit hours
Engl, Composition, 3 credits
Chem, General Chemistry, 4 credits
Math, Analytic Geometry, 3 credits
Math, Calculus I, 3 credits
ME, Elementary Forge and Machine Shop, 1 credit
IE, Engineering Problems, 2 credits
GE, Freshman Orientation, 1 credit
ROTC, Military or Air Science, 1 credit

Sophomore
Fall, 18 credit hours
Phys, Engineering Physics, 3 credits
Phys, Physics Lab, 1 credit
Math, Calculus II, 3 credits
Econ, Principles of Economics, 3 credits
IE, Descriptive Geometry, 2 credits
ME, Pattern Shop, 2 credits
A&S, Humanities, 3 credits
ROTC, Military or Air Science, 1 credit

Spring, 18 credit hours
Phys, Engineering Physics, 3 credits
Phys, Physics Lab, 1 credit
Math, Calculus III, 3 credits

Econ, Principles of Economics, 3 credits
EM, Statics, 3 credits
IE, Engineering Drawing II, 2 credits
ME, Machine Shop, 2 credits
ROTC, Military or Air Science, 1 credit

Junior
Fall, 19 credit hours
EM, Dynamics, 3 credits
ME, Thermodynamics, 3 credits
ME, Mechanism, 2 credits
Spch, Principles of Effective Speaking, 2 credits
ME, Seminar, no credits
EE, Electrical Circuits and Machines, 3 credits
Humanities Electives, 6 credits

Spring, 18 credit hours
EM, Mechanics of Materials, 3 credits
ME, Thermodynamics II, 3 credits
ME, Kinematics, 3 credits
ME, Mechanical Lab, 2 credits
ME, Seminar, 1 credit
ME, Heat Transfer, 3 credits
EE, Electronics, 2 credits
EE, Electrical Equipment Lab, 1 credit
ME, Inspection Trip, no credit

Senior
Fall, 18 credit hours
ME, Steam Power Plants, 3 credits
ME, Machine Design I, 3 credits
ME, Advanced Mechanical Lab, 2 credits
EM, Mechanics of Fluids, 3 credits
ME, Seminar, no credits
EM, Mechanical Fluids Lab, 1 credit
ME, Internal Combustion Engines, 3 credits
Humanities Electives, 3 credits

Spring, 17 credit hours
Phys, Elements of Atomic Physics, 3 credits
ME, Machine Design II, 2 credits
ME, Advanced Mechanical Lab, 2 credits
ME, Metals, 3 credits
ME, Seminar, 1 credit
Technical Electives, 6 credits[5]

By the end of the Jefferson years, seminars, electives, and humanities requirements had been adjusted. Students took seminars in the second semester of the junior and senior years rather than all year. The credits required during the junior and senior years now were more even than during the 1958–1959 school year. The department provided seniors with more options as the requirements allowed for nine hours of mechanical engineering electives.

The department set up its own course options for mechanical engineering students and required them to take a course in Western Civilization called "Institutions and Ideas of Western Man." During the first semester of the junior year, students had to choose either a world literature class "Introduction to Literature," or a philosophy class "Introduction to Philosophy." Also, students selected nine additional hours from eleven different departments in the College of Arts and Sciences in addition to these other courses.[6]

Curriculum, 1968

Freshman
Fall, 16 credit hours
Engl, Composition, 3 credits
Chem, General Chemistry, 4 credits
Math, Calculus I, 5 credits
GE, Engineering Graphics I, 2 credits
ME, Hot Forming, 1 credit
GE, Engineering Orientation, no credits
ROTC, Military or Aerospace, 1 credit

Spring, 17 credit hours

Engl, Composition, 3 credits
Chem, General Chemistry, 4 credits
Math, Calculus II, 5 credits
ME, Machine Process and Forming, 1 credit
GE, Engineering Methods, 1 credit
Spch, Principles of Effective Speaking, 2 credits
ROTC, Military or Aerospace, 1 credit[7]

Sophomore
Fall, 17 credit hours

Phys, Engineering Physics I, 3 credits
Phys, Physics Lab I, 1 credit
Math, Calculus III, 3 credits
GE, Engineering Graphics II, 2 credits
ME, Metallurgy and Materials, 3 credits
WCiv, Institutions and Ideas of Western Man, 3 credits
IE, Introduction to Computers, 1 credit
ROTC, Military or Aerospace, 1 credit

Spring, 17 credit hours

Phys, Engineering Physics I, 3 credits
Phys, Physics Lab II, 1 credit
Math, Differential Equations, 3 credits
ES, Statics, 3 credits
ME, Mechanism, 3 credits
WCiv, Institutions and Ideas of Western Man, 3 credits
ROTC, Military or Aerospace, 1 credit

Junior
Fall, 18 credit hours

Phys, Modern Physics III, 3 credits
ES, Dynamics, 3 credits
ES, Mechanics of Materials, 3 credits
ME, Thermodynamics, 3 credits
ME, Materials and Heat Treat Lab, 3 credits
WLit, Introduction to Literature, or
Phil, Introduction to Philosophy, 3 credits

Spring, 19 credit hours
ME, Thermodynamics, 3 credits
ME, Dynamics of Machinery, 3 credits
ME, Mechanical Engineering Lab, 2 credits
ES, Mechanics of Fluids, 3 credits
ES, Mechanics of Fluids Lab, 1 credit
EE, Electric Circuits and Machines, 3 credits
ME, Seminar, 1 credit
ME, Inspection Trip, no credits
Econ, Principles of Economics, 3 credits

Senior
Fall, 19 credit hours
ME, Machine Design, 3 credits
ME, Mechanical Engineering Lab, 2 credits
EE, Electronics, 2 credits
ME, Electives, 3 credits
ME, Heat Transfer, 3 credits
Technical Elective, 3 credits
Humanistic-Social Studies, 3 credits

Spring, 18 credit hours
ME, Design Project, 2 credits
ME, Mechanical Engineering Lab, 2 credits
ME, Elective, 6 credits
EE, Electrical Equipment Lab, 1 credit
ME, Seminar, 1 credit
Technical Elective, 3 credits
Humanistic-Social Studies Elective, 3 credits[8]

Pi Tau Sigma and the Society of Automotive Engineers Chartered
Pi Tau Sigma, the national honorary fraternity for mechanical engineers, had arrived on campus by the spring of 1959, largely attributed to the efforts of Professors Thomas B. Jefferson and David O. Watts. The national president Carroll M. Leonard and the secretary-treasurer David S. Clark installed the local chapter on May 8, and the chapter inducted its first pledge class on December 11, 1959. Pi Tau Sigma members proved the

Pi Tau Sigma officers. (*Razorback*, 1960)

chapter did not exist for the sole purpose of high scholarship, though. The chapter sponsored a student-faculty bowling meet for those in the Mechanical Engineering Department, and the event went well enough to plan a meet regularly every semester. They also started an engraving service for engineering students' slide rules as a fundraiser. Students responded favorably to the idea of personalizing their constant companions, and the organization benefited quite a bit financially from the endeavor. The chapter also provided assistance as guides for the mechanical engineering lab during open house.[9]

In November of 1963, the Society of Automotive Engineers (SAE) granted students a charter. According to the first president, membership "numbered less than fifteen. We shared an interest in automobile and aircraft, so SAE seemed to perk our career interests more than ASME. We were a small group, but did many events like judging car shows, working projects with the local high school to expose students to engineering, and championing the derby day races." Although mechanical engineering students usually had the most interest in such an organization, other engineering students could join as well. In the spring of 1966, the group sponsored an air force-aerospace exhibit. In the fall of 1966, the organization hosted the first annual Industrial Exposition to introduce students to engineering careers. The hard work put forth organizing the two events paid off when, in the spring of 1967, the national SAE presented $100 to the chapter for being named the most outstanding in the nation in the classification with 25–49 members.[10]

ASME continued to have a high participation as well. In the spring of 1961, the College of Engineering awarded the students in ASME a $20 prize for the excellent display set up for open house. At the regional spring conference in April, J. C. Hale and J. L. Mosley received awards for their presentations, and the chapter received $15 for placing second in attendance. In December of 1963, two ASME members, Dicky Bushmiaer and Ray Strange, faired well at the regional conference in Tulsa. Bushmiaer won $35 for his second-place paper on "Coming Launch Operations,"

SAE officers. (*Razorback*, 1967)

and Strange won $40 for his first-place paper on "Demineralizing Feedwater." In the spring of 1964, members decided to change officer terms from September through September to February through January. They thought the change would allow officers to become more familiar with their duties.[11]

A student remembered organizing the ASME open house during Engineer's Week in 1963, and that "student chapter members ran selected lab test equipment, engine dynamometers, steam engines, etc. all under the careful eyes of Professors Cogburn and Gleason. One of those professors arranged for a local auto repair garage to loan the ME Department a portable chassis dynamometer to set up on the faculty parking lot just outside the ME lab. To add a little excitement to the open house, I removed

ASME members. (*Razorback*, 1964)

the mufflers from my 1963 Ford and ran several full throttle demonstra-
tions of Detroit V8 power. It made a lot of noise and attracted a lot of visitors." Apparently, a rivalry developed during the day between the Ford
and another mechanical engineering student's 1963 turbo-charged Corvair
Spyder, and a drag race in Tulsa at an ASME regional conference later that
spring ultimately decided the fastest vehicle.[12]

Construction of the Mechanical Engineering Building

By the late 1950s, the increased enrollment and crowded conditions in the
College of Engineering prompted a study that recommended floor space
two to three times the size of the 91,042 square feet the college had at the
time. The Arkansas Legislature acted on the study in the 1961 special session by appropriating funds for an engineering-science center, which would
be made up of three buildings and cost $2 million. The National Science
Foundation assisted by providing grants to ease the state's expense. The
largest of the three buildings eventually housed the electrical engineering,
mathematics, and zoology departments in 100,000 square feet and seven
stories. The Mechanical Engineering Building occupied the 30,000-square-
foot building across Dickson Street from the new Science-Engineering
Center, and the third building contained a 500-seat auditorium.[13]

The new Mechanical Engineering Building, part of it built around the
old civil engineering laboratory, doubled the department's space. The old
highway research lab had solid concrete walls to withstand the testing of
various materials, and those walls remain part of the building today.

The main floor plan included the departmental offices, classrooms, a

Architectural plans for the Science-Engineering Center, ca. 1961. The Mechanical Engineering building is the background. (Mullins Library Special Collections, photo collection #1294)

small drafting room, and research and teaching labs in fuels and combustion, refrigeration, and materials and heat-treating. The basement plans included a thermo lab with available steam, hot and cold water, natural gas, AC/DC power, and compressed air. The lab also contained seven pads to mount engines and absorb vibrations. The basement also included a Lycoming aircraft engine in a separate test cell, and students in the design project class completed the engine test setup for the cell. The main hall in the basement provided enough room for a truck to move equipment. The only labs moved from the mechanical engineering shop building were materials and heat-treating and photo-micrographic.[14]

Mechanical Engineering building under construction. (*Arkansas Engineer*, March 1966)

Front porch of the Mechanical Engineering Building, ca. 1980. (Mullins Library Special Collections, photo collection #4725)

Even though the new lab had better equipment than the old lab, the possibility of human error remained. A MSME student instructing a lab in the late 1960s explained that an engine experiment with a "cooling system was a once-through tap water supplied system, manually controlled. My senior engineering students cracked up when the engine turned into a steam generator because I started them with the water at a low rate, then forgot to tell them to raise the rate as the load increased. Hot water and steam blew all over the place, but the engine survived and none of the faculty ever found out."[15]

With its new facility, the Mechanical Engineering Department could begin to offer more focused programs, such as nuclear engineering, and provide more laboratory space for an increasingly intellectually sophisticated student body, as evidenced by the chartering of Pi Tau Sigma and the SAE. Although there would be less activity in the Mechanical Engineering Department in the 1970s than in the 1960s, nuclear engineering work would continue into the next decade as well as social changes in general.

The College of Engineering had expanded out of Engineering Hall into the Science-Engineering Center by the 1963–1964 academic year. The Electrical and Mechanical Engineering Departments completely moved into the new buildings, while industrial engineering moved only its graphics work from Engineering Hall. The space left after these departments relocated allowed the engineering mechanics, industrial, civil, and chemical engineering departments to spread out of crowded conditions that had existed for some time. The lack of space had been an issue in attracting not only research projects but students as well, so the new buildings had an immediate impact as the college hired seven faculty members with PhDs and received more research grants over the course of the 1964–1965 academic year.[16]

The College of Engineering during the Period

In the fall of 1959, the College of Engineering began offering classes leading to the degree of doctor of philosophy after the university had approved the curriculum during the summer, perhaps a contributing factor to the increase in graduate enrollment. In the College of Engineering, the enrollment figures paralleled the national trends in the early 1960s as undergraduate enrollment declined, but graduate enrollment increased. The increase in graduate students and faculty resulted in more research than ever before with expenditures increasing from $4,874,903 in 1960 to $15,456,794 by 1970, but that amount still did not equal the support research received in many other states.[17]

Engineering students expressed concerns over draft regulations that required students to complete a certain number of credit hours toward their degree each academic year to receive a draft deferment. Since an engineering degree could sometimes take as long as five years to complete, this meant students would have to hope for a special deferment or National Guard service if they stayed in school longer than eight semesters. An editorial in the *Arkansas Engineer* advocated the university allowing engineering degrees to be five-year degrees so students could have more productive experiences without consternation over the draft.[18]

The concern over the draft probably impacted engineering enrollment as engineering students represented a lower proportion of the total student population even as undergraduate enrollment increased at the university in the early 1960s. Dean George Branigan found this to be a surprising statistic in light of the fact that engineering students usually commanded the highest salaries coming out of college, but he speculated the courses might have become too rigorous possibly because of inadequate preparation in high school. Nevertheless, expansion of facilities did not slow down as enrollment declined because the college had to have the necessary equipment and space to attract more students. The laboratory areas and equipment had to be more modern to compete for students, as well as faculty, many of which received much lower salaries compared with other universities and the private sector. The salary discrepancy meant that the college had fewer faculty members with PhDs than they desired but hoped the new facilities and increased research opportunities would compensate for the salary discrepancies.[19]

Each engineering department examined its requirements and made several adjustments in the early 1960s. In order to develop more highly skilled graduates, the college began to place students into higher-level mathematics courses in their freshman year. The committee also decided to increase the number of humanities credits to eighteen hours to develop more well-rounded students, and every department, except for mechanical and industrial engineering, eliminated shop courses. All the departments reduced credit hours in engineering graphics. The committee hoped the changes would result in students with greater skills in fundamental engineering along with a stronger humanities foundation in all of the departments.[20]

The college also changed the mission of the Engineering Mechanics Department to facilitate more science and technology instruction. In 1966, Engineering Mechanics became Engineering Science in an attempt to expand beyond the courses taught in engineering mechanics such as statics, dynamics, and elasticity and to focus more on science and less on structural engineering. In order to focus more on undergraduate instruction, the college began offering a bachelor of science in engineering science (BSES), whereas this type of degree had been unavailable through the Engineering Mechanics Department.[21]

Perhaps as a way to encourage better writing skills among engineering students, an award was established in 1968 by the will of Anna Powell Wood to recognize outstanding writing in the field of engineering. Her husband, Albert G. Wood, received one of the first BME degrees in 1892 and an honorary doctor of science degree in 1942. A mechanical engineering student, Lawrence Dyer, BSME (1968), won the first award for his paper "A Survey of Nuclear Rocket Systems."[22]

St. Patricia skits during the Engineer Rally in the 1960s remained "profane and vulgar" despite the pleas from the *Arkansas Engineer* editorial page. In an attempt to discourage racy exhibitions, the editor Bob Wilson pointed out that the most outrageous skits did not always win the competition. However, the skits continued to showcase questionable material and females in more revealing outfits. By the spring of 1965, Dean Branigan had joined the editor of the *Arkansas Engineer* in condemning the subject matter of the skits. Both indicated that unless changes occurred, the rally would not happen again in the future. Perhaps the statements appeared too late to make a program change, or the students simply did not heed the warnings, because one skit during the rally featured women in Playboy bunny outfits. By the early 1970s, the annual celebration had problems even attracting a decent crowd as an editorial in the *Arkansas Engineer* requested more student participation during Engineer's Week. The dean pointed out that the low attendance did not warrant the money spent on the activities.[23]

The University during the Period

The University of Arkansas campus reflected not only the national trend of university expansion, but also the expansion of thought and issues and the controversy that accompanied both in the 1960s. After Dr. Albert Ellis presented a lecture that apparently advocated premarital sex to a packed crowd of students on the university campus in February of 1962, university administrators decided to closely monitor future speakers to avoid the turmoil that surrounded Ellis's comments. By 1965, the university had adjusted the policy due to criticism from both students and faculty who claimed the Facilities Use Committee scrutinized the speakers so much that it amounted to blatant censorship.[24]

Changes in creative writing also resulted in confrontation among students, faculty, and administrators. In the spring of 1966, the literary magazine *Preview* sent an issue to be printed at the university press, but the printer hesitated doing so because he deemed some of the language to be too obscene. After consulting with his superiors, they agreed the issue should not be printed. The Arkansas Association of University Professors (AAUP) argued that refusal to print the magazine constituted censorship, and students demonstrated in front of President David Mullins's office. Finally, in February of 1968, the university arranged for the printing of the magazines with private money, and copies were distributed to the student body.[25]

During the 1960s, the University of Arkansas was placed on the American Association of University Professors' censure list after a series of events that began in 1958, the year the Arkansas General Assembly passed Act 10. The act mandated that all public schoolteachers and other educators supported by public funds submit affidavits with information regarding the employees' involvement with outside organizations, such as those that might be sympathetic to integration of public schools. University faculty, President John T. Caldwell, and the board of trustees all thought the act would unduly interfere with personal liberties, and injunctions against submitting such information were obtained until lawsuits against the act could be settled. After the lower federal courts and the Arkansas Supreme Court deemed the act constitutional, the university requested the affidavits, and all but four faculty members submitted them. The board of trustees decided the faculty members no longer worked for the university as a result. Eventually, the United States Supreme Court struck down the law in 1960, but the university did not reinstate the four faculty members who refused to submit affidavits. This refusal to acknowledge the displaced faculty members led to the AAUP censure in 1964, which lasted for four years, until the university made amends and offered concessions to the four in 1968. According to Thomas Jefferson, the Mechanical Engineering Department head at the time, "I do not recall [Act 10] having a disastrous effect on the College of Engineering, though I doubt anybody thought it was a good piece of legislation. Most of us were used to security checks by the federal government and familiar with the procedure of listing all organizations we belonged to . . . I am not aware of any engineering people that did not go ahead and submit the required information."[26]

The University of Arkansas experienced unprecedented growth in the 1960s and hired more faculty and staff in addition to the construction of new buildings. By 1970, 608 faculty and staff worked at the university compared with 358 in 1960. Undergraduate students doubled from 5,862 in 1960 to 11,709 by 1970, and graduate enrollment tripled from 501 in 1960 to 1,621 in 1970.

In order to handle the increased student enrollment in the 1960s and future growth, President Mullins pushed for a massive series of construction projects for classrooms, dormitories, and administrative offices. The university favored high-rises for three of the residence halls, Yocum, Humphreys, and Hotz, and shorter, more expansive designs for Fulbright, Pomfret, and the Carlson Terrace expansion. The large academic buildings spread over even larger areas than the residence halls. Mullins Library, the Graduate Education Building, the Science-Engineering Building, and additions to the Animal Science Building radically changed the appearance of the campus, as did the student union and the open mall between it and the library. University administration moved to a new building on Maple, and the Physical Plant shifted its facilities to an area southwest of the campus. Hundreds of thousands of dollars from private donors helped to finance the construction and to set up much-needed scholarships and endowments to attract students and faculty members to conduct research in the new facilities.[27]

Student groups had organized by the late 1960s to protest probably the two most important issues at the time, the Vietnam War and civil rights. The Southern Students Organizing Committee (SSOC) protested against Dow Chemical when the company recruited on campus in 1968. Dow Chemical had similar problems on other campuses as a result of its manufacture of napalm used in the Vietnam War. Although the protest remained calm compared with others around the country, the company officials indicated they interviewed fewer students than they had anticipated. Some members of SSOC also associated with Black Americans for Democracy (BAD), a group that formed shortly after the assassination of Dr. Martin Luther King Jr. in April of 1968. The organization questioned certain policies on campus and targeted many for change, such as the lack of black faculty and staff, the omission of black history courses, and the association of the confederate battle flag and the song "Dixie" with athletic events.[28]

Although other land-grant institutions around the country had ended compulsory military drill by the late 1960s, the university still required students to participate. In 1965, preliminary studies began and by the spring of 1969, a committee recommended changing military art courses to voluntary. Some discussion had ensued about whether ROTC should remain a requirement, but that program changed to voluntary as well. Although enrollment in freshman and sophomore military art courses had dropped by 1970, students remained interested in officer training, which seemed to mesh well with the changing sophistication of the United States military as it depended less and less on conscripted ground troops and more on highly trained volunteers, such as those in aerospace studies.[29]

The town of Fayetteville spread out as subdivisions and shopping centers began to dot the landscape to accommodate the population increase from 20,000 to 30,000 residents by the end of the 1960s. The new residents included former college students, retirees, and others attracted to the small-town atmosphere of Fayetteville. In addition to the development of new residential and shopping areas, town business leaders sought to ensure that the square remained a viable commercial area and secured land to attract light industry. However, the largest employer in the 1960s had become and would remain the university, and shops and restaurants along Dickson and elsewhere would benefit from the business of nearly 12,000 students by the early 1970s.[30] Those students also helped to fill up the new academic and residence buildings that had been the focus of so much construction in the 1960s.

Athletics

The football program provided escape from the controversies on campus as Frank Broyles led the Razorbacks to one successful season after another in the 1960s. The teams went to three straight post-season bowl games in the early 1960s before ending the 1963 season at 5-5 due to a young team. However, the team rebounded next year and completed a prefect season, and polls taken after the bowl games made the Razorbacks national champions. Many of the same players returned the next year, and the team finished with a perfect regular season and would have repeated as national champions had they not lost the Cotton Bowl to Louisiana State University 14-7. By the 1967 season, the team experienced an average record as starters graduated, but the team came back again over the next two years. The 1969 season featured the nationally televised "Great Shootout" between Arkansas and Texas that Arkansas lost 15-14 as President Richard Nixon watched from the stands in Fayetteville. The respect Frank Broyles and the football program commanded during the 1960s could be evidenced by the many assistant coaches who left during that time to accept head coaching positions and by the long list of players who became All-Americans.[31]

Biographical Sketches

Thomas B. Jefferson, BSME (1949), Kansas State University; MSME (1950), University of Nebraska; PhD (1955), Purdue University; Professor and Department Head (1958–1969)

After receiving a BSME from Kansas State, Thomas B. Jefferson taught classes at the University of Nebraska for two years while working toward an MSME degree. He left in 1952 to teach and work on his PhD at Purdue. He taught at Purdue after receiving his degree there in 1955 until coming to the University of Arkansas where he taught courses that included Thermodynamics, Heat Transfer, and Power Plants. He became associate dean of the College of Engineering and associate director of the Engineering Experiment Station in 1968 before accepting a position at

Thomas B. Jefferson, ca. 1969. (Mechanical Engineering collection)

Southern Illinois University as the dean of the School of Technology that became effective July 1, 1969. Ken Deaver, who had been assisting Jefferson with administrative duties in the department, became department head.[32]

As department head, Thomas Jefferson often conferred with workers as problems arose during construction of the new Mechanical Engineering Building. Most were solved with little difficulty, but one in particular almost led to a serious delay. After the foundations and supports had been completed and the walls had begun to be attended to, Jefferson received a call from a physical plant employee who noticed an incorrect measurement at the front entrance. Two columns "were four inches further apart than they should have been. And it was not a symmetrical error; all four of the extra inches were on one side of the center line. Wow! To correct this would have required breaking out all of the work that had been done so far. We walked around and considered the delay . . . and said, 'Who would ever notice in this large a space?' So we decided to let it go." Even after the building was completed, some small adjustments had to be made, such as the glass in the men's restroom that had been installed backward allowing those in the building to see into the restroom.[33]

Donald A. Gilbrech, *BSIE, MS, University of Arkansas; PhD, Purdue University; Professor (1980–1990)*

Donald A. Gilbrech, ca. 1981. (Mechanical Engineering collection)

Dr. Donald Gilbrech was born in 1927, graduated from the University of Arkansas in 1953, received his master's degree in 1954 from the University of Arkansas, and completed his PhD at Purdue University in 1958.[34] After many years, beginning in 1953, in the Engineering Mechanics Department and later in the Engineering Science and Mechanical Engineering Departments, Dr. Gilbrech retired in 1990.

Gilbrech headed research projects in fluid flow for the National Science Foundation and for the National Institutes for Health. He did consulting in the areas of fluid flow, high-speed photography, and electrical sensors. He was president of Applied Mechanics, Inc., for twelve years, developing and manufacturing laboratory equipment for colleges and universities.

He acquired patents on a variety of devices and was a registered professional engineer in the state of Arkansas. He taught computer interfacing, supervised the mechanical engineering laboratory courses, and helped to modernize the mechanical engineering laboratories. His hobbies were playing banjo, woodworking, and tennis.

James R. Kimzey, BSME, University of Arkansas; MSME, PhD, Kansas State University; Associate Professor (1980–1982)[35]

Dr. James Kimzey is a native of Arkansas and a 1961 BSME graduate of the University of Arkansas. His MS and PhD (1966) in Applied Mechanics are from Kansas State University. As an officer in the U.S. Army, Jim served as military detailee to the Marshall Space Flight Center in the areas of dynamics and control of spacecraft (Skylab). Jim returned to Fayetteville in 1968 to teach in the Engineering Science Department and later in the Mechanical Engineering Department. As a faculty member for fourteen years, his interests were solid mechanics, numerical methods, and machine design. He was also heavily involved as faculty advisor to engineering groups, such as in the mini-baja contests and the ASME. In addition to hands-on teaching methods, his practical applications work involved business and industry, as well as community service.

In 1982, Kimzey joined Baldor Electric Company, manufacturers of industrial electric motors, and became vice president for research and engineering. He has been central to the entry of Baldor into the servo motor and control market in the United States and Europe. His work continues into research and design of microprocessor-based servo-motion controls,

James R. Kimzey, ca. 1981. (Mechanical Engineering collection)

variable speed industrial products, research and testing facilities for new products, and interdisciplinary coordination of engineering and business functions.

Henry H. Hicks Jr., BSChE (1943), University of Arkansas; MSE (1948), PhD (1960), University of Michigan; Professor (1961–1991)

Professor Henry Hicks taught Thermodynamics and Heat Transfer. In addition to work with the military on ammunitions, Hicks was also involved in researching municipal problems with garbage disposal. He also consulted insurance companies concerning accidents with fire.[36]

A student from the late 1960s recalled that "Hicks was a character and would cut up a little sometimes and was always in loose, wrinkled clothes like the TV detective Colombo. Dr. Jefferson would leave his pipe in the hall on the ashtray of fire extinguishers when he went into a classroom. A couple of times, I saw Hicks move the pipe a little ways so that Jefferson would have to search for it after class. He did consulting work outside of the university and had a lot of tales about those jobs and projects. The one

Henry H. Hicks Jr., ca. 1981. (Mechanical Engineering collection)

that comes to mind was an assignment to investigate 'ghost lights' some-where in Missouri. He had stories of mysterious lights along railroads, highways, and hillsides, some he could explain and some he could not. There was always a group of us that would stay and listen when he told one of those stories. A lot of us regarded Hicks as a real engineer that did things in the real world whereas some of the other professors, particularly those outside of our department, were regarded as 'ivory tower' types who taught but did not have experiences like Hicks did."[37]

Another student from around the same time also enjoyed Hicks's unconventional teaching style, recalling that "his tests were open book, open notes, for as he liked to say, 'They will not do you any good anyway.' I remember the first day of Thermo I when he was explaining his grading system. He said there would be four tests worth 10 percent each, three pop quizzes worth a total of 20 percent, and a final exam worth 25 percent. One of the students piped up, 'But Professor Hicks, that adds up to only 85 percent.' Professor Hicks walked over to his desk, bent down so that their noses were about six inches apart, and said, 'The other 15 percent is personal evaluation.' Although he did not fit the mold of most people's idea of a college professor, I am glad that I had the experience, and I feel that it has been invaluable to my career."[38]

Helmut Wolf, *BSME (1948), Case Institute of Technology; MSME (1950), PhD (1958), Purdue University; Associate Professor (1961–1966), Professor (1966–1972), Distinguished Professor (1972–1988), Interim Department Head (1984–1985)*

During his time at the university, Helmut Wolf was involved with numerous publications, including his own textbook, *Heat Transfer,* in 1982. Although he worked on a number of consulting and research projects, those activities did not interfere with his classroom presence, as the instructor awards he won proved. In the 1960s, many of the MSME students worked under the direction of Wolf, and he and Professor Jong worked to integrate the Mechanical Engineering and Engineering Science Departments in 1980. Courses he taught included heat transfer and thermodynamics. Wolf received the first Raymond F. Giffels Professorship in Engineering provided in the spring of 1971 by Mrs. Erma Fitch Giffels.[39]

Helmut Wolf, ca. 1981. (Mechanical Engineering collection)

According to a student from the early 1960s, Wolf "was noticeably the most competent of the staff and had a teaching manner that made it easy to learn. He was well respected and approachable—always willing to share his knowledge." Another student from that same time recalled his first encounter with Wolf "at the registration table as a very intimidated freshman, coming out of the hills of Arkansas into the 'big university' world. I had very high test scores in math, and he wanted me to go directly into calculus. I insisted that I was not ready, and he became animated and crazed and frustrated towards me. 'How can we stay ahead of the Russians?' He threw up his hands and let me take advanced math first. It turns out he was half right—I aced all the math and could have gone directly to calculus—but he was wrong about it allowing the Russians to get ahead of us."[40]

Jim H. Akin, BSME (1960), MSME (1961), PhD (1967), University of Texas; Assistant Professor (1963–1968), Associate Professor (1968–1974), Professor and Head of the Department of Engineering Science (1974–1980), Professor (1980–1998)

Classes that Professor Jim Akin taught included Mechanisms, Dynamics, Mechanical Design, Kinematics, and the senior design projects. In the early 1970s, Akin worked with several other professors to create a block of technical studies for students interested in system dynamics. He showed a real interest in students by serving as faculty advisor for the local chapter of Theta Tau in the 1960s, then advising the student staff of the *Arkansas Engineer*.[41]

A student from the early 1960s recalled that when Akin first started, "he was the youngest professor on the ME staff and probably related to us fledgling engineers best due to the small age difference. Jim graduated from the University of Texas and hence we always referred to him as 'Texas Jim.' I always remember Jim as an affable, outgoing individual, well liked by students."[42]

A student from the late 1960s had fond memories of Akin as well, recalling that as he would lecture, he would "write equations with his right hand, and, as he moved across the blackboard, he would simultaneously use a long handle dust mop to erase the stuff behind him. I probably learned more about problem solving from Akin than anyone else in the college of engineering." The former student also recalled the good rapport Akin had with the students outside of the classroom, such as when he

Jim H. Akin, ca. 1981. (Mechanical Engineering collection)

would "sit in the break room sometimes between classes, talking and joking with us. The first Pizza Hut to open in Fayetteville was on North College. It had a basement, a jukebox, and very cold beer on tap. I do not remember what the occasion was, but I do remember Dr. Akin sharing beer and pizza with a group of ME seniors with the Beatles singing 'Hey Jude' over the jukebox. When the student ASME section had a couple of kegs of beer on a White River gravel bar east of Fayetteville, Akin was there with us, and he may have even picked up part of the bill."[43]

Ing-Chang Jong, BSCE (1961), National Taiwan University; MSCE (1963), South Dakota School of Mines and Technology; PhD (1965), Northwestern University; Professor (1980–present)[44]

Dr. Ing-Chang Jong taught in two different departments before his affiliation with the Mechanical Engineering Department. He was first hired by the Engineering Mechanics Department in 1965. The next year, that department was renamed Engineering Science. In 1980, that department was merged with Mechanical Engineering.

He was one of the first engineering professors to receive the competitive grants that had begun to be offered by the late 1960s. From 1967 to 1969, he worked on a National Science Foundation Research Initiation Grant. The same year that grant ended, he received another one from the National Science Foundation from 1969 to 1971. During this time, he also worked hard in the classroom, winning Outstanding Educator of America in 1971 and initiating the teaching of Tensor Analysis and Continuum Mechanics in 1967.[45]

Ing-Chang Jong, ca. 1966. (Mechanical Engineering collection)

Around 1990, he focused a lot on the concept of "displacement center" in statics, which made possible the use of just geometry and algebra (rather than differential calculus) as prerequisite mathematics for, and allowed great expansion of, the use of the "principle of virtual work" in statics and dynamics. This work resulted in the publication of several books published with B. G. Rogers that included *Engineering Mechanics: Statics* (1991), *Engineering Mechanics: Dynamics* (1991), and *Engineering Mechanics: Statics and Dynamics* (1991). He also edited, with F. A. Akl, *Developments in Theoretical and Applied Mechanics, Volume XVII* (1994).[46]

His long career at the university resulted in his teaching the children of former students. He recalled "one event among others that I couldn't forget. Van Parker (an EE student in my Statics class in 1967 who later became a dentist) came to sit with his son, Grant Parker (a ChE student at the time), in my 7:30 a.m. summer session class in Statics on Friday, June 14, 2002, in ME 212. It was about five minutes into my own lecture before I realized that the father and son were sitting next to each other in the class the son was taking. Immediately, I recognized the special visitor and welcomed him to the class. The class was obviously surprised by the event. Nevertheless, the lecture and discussion continued. After the class, Van told me, 'I can still remember the same Statics topics and concepts you taught in the class I took thirty-five years ago.' I was touched that the father took the time to join his son in revisiting the class he took long ago from the same professor."[47]

Chapter 5

The Deaver Years, 1969–1980

The faculty that had been hired during Paddock's later years and Jefferson's time continued to provide a solid base of instruction. Several of the faculty hired during the 1970s did remain for a fair amount of time, which provided the department with a high level of stability. Also, all of them had either a PhD or equivalent, signaling a new standard of preparedness for faculty that had developed under the dean of the College of Engineering, George Branigan.

Department Head
Franklin Kennedy (Ken) Deaver, 1969–1980

Faculty
1. Jack H. Cole, BSME, MSME, PhD, Oklahoma State University; Assistant Professor (1968–1972), Associate Professor (1972–1977), Professor (1977–1982, 1994–2002)
2. James R. Kimzey, BSME, University of Arkansas; MSME, PhD, Kansas State University; Assistant Professor (1968–1969), Associate Professor (1980–1985)[1]
3. Pradeep B. Deshpande, BSC, Karnatak University, India; BSChE, MSChE, University of Alabama; PhD, University of Arkansas; Assistant Professor (1969–1970)
4. John Lawrence Head, BSc, DIC, PhD, University of London; Visiting Associate Professor (1972–1973)
5. Milton E. McClain Jr., BA, Emory University; MS, University of Idaho; PhD, Georgia Institute of Technology; Associate Professor (1974–1982)
6. Arvid Myklebust, BSME, MSE, PhD, University of Florida; Assistant Professor (1975–1981)
7. William Danny Turner, BSME, MSME, University of Texas; PhD, University of Oklahoma; Associate Professor (1976–1982)
8. Charles M. North Jr., BAeroEngr, MS, University of Florida; MA, PhD, University of Alabama; Associate Professor (1971–1980)

9. O. Hugo Stordahl, Siv.ing., Dr.ing., University of Trondheim; Visiting Associate Professor (1978–1980)

By 1969, the university ceased to require students to take either military arts or aerospace education, but students could still receive credit if they chose to enroll in courses offered. The department no longer required inspection trips during the junior year. However, a similar experience could be gained through the cooperative education program. Two students who had completed their sophomore years could choose to alternate every six months working in an industry of their choice. One student would work while the other remained in school, and then they would switch until the completion of their degrees. Also, semester hours had been reduced for every class. The course requirements for the senior year allowed students to tailor the schedule to their own interests as many of the credit hours consisted of electives.[2]

Curriculum, 1969

Freshman
Fall, 15 credit hours
Engl, Composition, 3 credits
Chem, College Chemistry, 5 credits
Math, Calculus I, 5 credits
GE, Engineering Graphics I, 2 credits
GE, Engineering Orientation, no credit

Spring, 16 credit hours
Engl, Composition, 3 credits
Phys, University Physics I, 3 credits
Phys, University Physics Lab, 1 credit
Math, Calculus II, 5 credits
ME, Industrial Processes, 2 credits
GE, Engineering Graphics II, 2 credits

Sophomore
Fall, 16 credit hours
Phys, University Physics II, 3 credits

Phys, University Physics Lab II, 1 credit
Math, Calculus III, 3 credits
ME, Metallurgy and Materials, 3 credits
Econ, Principles of Economics, 3 credits

Spring, 16 credit hours
Phys, University Physics III, 3 credits
Math, Differential Equations, 3 credits
ES, Statics, 3 credits
ME, Mechanism, 3 credits
IE, Introduction to Computers, 1 credit
WCiv, Institutions and Ideas of Western Man, 3 credits[3]

Junior
Fall, 18 credit hours
ME, Thermodynamics, 3 credits
ME, Materials and Heat Treatment Lab, 3 credits
ES, Dynamics, 3 credits
ES, Mechanics of Materials, 3 credits
EE, Electric Circuits and Machines, 3 credits
WLit, Introduction to Literature, or
Phil, Introduction to Philosophy, 3 credits

Spring, 18 credit hours
ME, Thermodynamics, 3 credits
ME, Dynamics of Machinery, 3 credits
ME, Machine Analysis, 3 credits
ME, Mechanical Engineering Lab, 2 credits
ES, Mechanics of Fluids, 3 credits
ES, Mechanics of Fluids Lab, 1 credit
EE, Electronics, 2 credits
ME, Seminar, 1 credit

Senior
Fall, 17 credit hours
ME, Machine Synthesis, 3 credits
ME, Heat Transfer, 3 credits

ME, Mechanical Engineering Lab, 2 credits
ME, Elective, 3 credits
Technical Elective, 3 credits
Humanistic-Social Studies Elective, 3 credits

Spring, 16 credit hours
ME, Mechanical Engineering Lab, 2 credits
ME, Elective, 6 credits
Elective, 4 credits
Humanistic-Social Studies Elective, 3 credits
ME, Seminar, 1 credit[4]

The course requirements for the BSME went through few adjustments under Deaver, but changes did occur with regard to the humanities requirements and the total hours required in some class years. Also, the department no longer required the junior and senior seminars by 1980.[5]

Curriculum, 1980

Freshman
Fall, 16 credit hours
Engl, Composition, 3 credits
Chem, College Chemistry, 5 credits
Math, Calculus I, 5 credits
GE, Engineering Graphics, 2 credits
GE, Engineering Orientation, 1 credit

Spring, 17 credit hours
Engl, Composition, 3 credits
Phys, University Physics I, 3 credits
Phys, University Physics Lab I, 1 credit
Math, Calculus II, 5 credits
ME, Industrial Process, 2 credits
Humanistic-Social Studies Elective, 3 credits

Sophomore
Fall, 16 credit hours
Phys, University Physics II, 3 credits

Phys, University Physics Lab II, 1 credit
Math, Calculus III, 3 credits
ME, Metallurgy and Materials, 3 credits
Econ, Principles of Economics I, 3 credits
Humanistic-Social Studies Elective, 3 credits

Spring, 16 credit hours
Phys, University Physics III, 3 credits
Math, Differential Equations, 3 credits
ES, Statics, 3 credits
ME, Mechanism, 3 credits
IE, Introduction to Computers, 1 credit
EE, Electric Circuits I, 3 credits

Junior
Fall, 15 credit hours
ME, Thermodynamics, 3 credits
ME, Materials and Heat Treat Lab, 3 credits
ES, Dynamics, 3 credits
ES, Mechanics of Materials, 3 credits
EE, Electronics I, 3 credits

Spring, 18 credit hours
ME, Thermodynamics II, 3 credits
ME, Dynamics of Machinery, 3 credits
ME, Machine Analysis, 3 credits
ME, Mechanical Engineering Lab, 2 credits
ES, Mechanics of Fluids, 3 credits
ES, Mechanics of Fluids Lab, 1 credit
Humanistic-Social Studies Elective, 3 credits[6]

Senior
Fall, 17 credit hours
ME, Heat Transfer, 3 credits
ME, Mechanical Engineering Lab, 2 credits
ME, Elective, 6 credits
Technical Elective, 3 credits
Humanistic-Social Studies Elective, 3 credits

Spring, 17 credit hours
ME, Mechanical Engineering Lab, 2 credits
ME, Machine Synthesis, 3 credits
ME, Elective, 6 credits
Elective, 3 credits
Humanistic-Social Studies Elective, 3 credits[7]

Mechanical engineering organizations remained strong with active members. The ASME and Pi Tau Sigma held some joint activities during the fall of 1972 including an outing to the 502 Club. In the fall of 1975, the ASME began sponsoring a trip to the spring regional conference for the winner of the local chapter project contest. The previous spring, Mike Hall won at the regional conference and presented his project at the national ASME convention. In the spring of 1977, the ASME planned for more events than they had in past years. The organization heard speakers from Alcoa, the Navy Nuclear Program, and Union Carbide. In addition, they planned field trips to Tulsa and Fort Smith and organized a picnic for the department.[8]

Research had always played a considerable role in the instruction of engineering, but funding had often been an issue. Research work expanded

ASME members. (*Razorback*, 1974)

Pi Tau Sigma members. (*Razorback*, 1976)

in the 1970s as funds became increasingly more available, particularly from the federal government. More than half of the 16,247 full-time faculty members nationwide worked on research projects totaling $288 million, and the federal government provided $220 million of that amount.[9]

The Southwest Experimental Fast Oxide Reactor (SEFOR) had been originally used to study safety aspects and shutdown procedures in the operation of a fast breeder reactor as concerns over the depletion of fossil fuels led to more experiments with nuclear power by the late 1960s. The facility for the $25 million program was built from 1965 to 1968, and workers performed tests there from 1968 to 1972. Bill Becker, the manager of SEFOR in 1971, explained that the plant produced almost no power since it was built for experiments as well as to prove nuclear power could be used to benefit society rather than destroy it as bombs dropped on Japan did to end World War II.[10]

After the government completed studies there in 1972, the facility, located twenty-five miles southwest of the university near Devil's Den, was offered to the university, and school officials accepted. The facility's main purpose was to serve as a calibration center, which was used mainly by the Mechanical Engineering Department. The department contracted with the National Bureau of Standards, which needed a method of accurately calibrating instruments used in nuclear plants for surveys of nuclear fields. Cogburn, who had returned from London with a PhD in nuclear engineering, served as the project director. One of the projects involved the loan of Californium –252 from the United States Department of Energy (DOE). The DOE loaned the SEFOR the two largest pieces of californium –252 outside of the DOE. The sources were installed in the reactor vessel in a big cell that had a manipulator, which allowed researchers to move things around externally. Experiments at the SEFOR used the element along with the D_2O sphere to calibrate neutron survey instruments.[11]

Eventually, the SEFOR cell developed into a good radiation source, and the National Bureau of Standards sent two people down to Fayetteville. The Bureau had previously checked the source in a large open space to determine its emission fields. In order to determine that SEFOR could also perform a similar calibration, the men tested the calibration at SEFOR based upon what they had found at their own testing facilities and determined that SEFOR could perform the proper calibration as well. Later in the early 1980s, research included the irradiation of different materials beside the outside of the vessel to learn what the radiation fields were that went through the vessel and might potentially cause embrittlement of the metal.[12]

Although SEFOR attracted concern as speculation over the dangerous effects of radiation arose, an experiment done there proved that fast breeder reactors could shut down safely. To control an unsatisfactory reaction, an element would be introduced into the reactor resulting in a super high heat to signal a shutdown. The reactor would then begin a slow shutdown process that could be accelerated by manual control. However, the irradiated steel around the vessel and the nickel reflectors would be lasting concern after experiments came to an end in the late 1980s. A series of locks placed on the surrounding fence, the outside of the building, and the inside refueling area left some with lingering concerns of possible accessibility, and funds began to be sought to make the area level with the ground to seal the reactor up permanently.[13]

The program in the Mechanical Engineering Department that had improved the most was obviously nuclear engineering. The benefits of having a facility so close to campus provided students and faculty with a unique opportunity for beneficial research and scholarship. As with facilities involving nuclear reactions nationwide, the safety of the SEFOR came into question, but the experiments conducted there more than likely helped with the program's continued success throughout the 1980s and 1990s.

By 1969, University of Arkansas BSME graduates averaged $827/month in starting salaries, second behind BSChE graduates who averaged $860/month. Of course, out-of-state salaries tended to be higher. Despite competitive salaries offered to recent graduates, mechanical engineering enrollment dropped from 196 students in 1971 to 145 by 1972. Other departments experienced similar declines as total college enrollment dropped from 994 to 846 during the same years. By the fall of 1975, though, the numbers in the college had improved from 785 in the previous fall to 935 in 1975. Also notable was the increased enrollment of women and blacks. Women represented a large enough group in the college that by the fall of 1975, they had organized the Society of Women Engineers (SWE). In 1973, black students had organized a chapter of the National Society of Black Engineers.[14]

Women and Minorities

Although the university found finances tight in the early 1970s, student aid continued to rise, reflecting not only the increasing costs of tuition, but also to support the higher enrollment of women and minorities. Women had pursued education in the field of engineering nationwide since the nineteenth century, but without sufficient opportunities, the numbers did not amount to a significant portion of the field as a whole. By the early 1970s, the number of women began to increase rapidly at institutions across the country and at the University of Arkansas. In the fall of 1971, eleven women enrolled in the college of engineering, a small number compared to the rest of the campus, but a 50 percent increase over the previous year, and the number had doubled again by the fall of 1972. The reasons varied for the increase, but there appeared to be a general sense that women talented in such subjects as math and physics in high school no longer felt limited to teaching because they were encouraged to seek out more technological careers. Women may have also been enrolling in larger numbers due to reports that the engineering field would have many more available jobs than applicants because of low numbers of engineering graduates as a whole. Qualified women appeared to have had an excellent chance to enter a male-dominated field in quite large numbers by the end of the 1970s. Two of the first women graduates from the University of Arkansas Mechanical Engineering Department were Naomi Suloway, BSME 1957, and Coralee Clifton, BSME 1958. After graduation, Suloway worked as a designer in the Chicago area, and Clifton worked for the Corps of Engineers in Little Rock.[15]

By the fall of 1975, enrollment picked up when 935 undergraduates began the year compared to 785 the year before. Women represented 6.6 percent of the total undergraduates, and blacks 2.2 percent. African Americans began to enroll in larger numbers not only in undergraduate courses but graduate as well. The university hoped to continue increased minority enrollment through students from the Pine Bluff campus in particular. Students there would complete two years of engineering studies then finish the rest of the degree at the Fayetteville campus. Although enrollment as a whole had increased, Dean Loren Heiple anticipated industry would make a genuine effort to make up for past deficiencies for the lack of minorities and women employed in industry.[16]

Indeed, the larger enrollment of women in engineering colleges spilled over into the workplace by the late 1970s. As in universities, the business world would accept women as equals more and more. Helen Bhardwaj, who graduated with a BSME, worked at TRW in Rogers in the early 1980s. In an interview with the *Arkansas Engineer,* she indicated that the only hardships she encountered concerned her inexperience in the workplace in general, which had nothing to do with her gender.[17]

Other minority groups, Hispanic, Asian, and Native American, enrolled in engineering courses in greater numbers nationwide as well in the 1970s. By the end of the decade, minority groups, including African Americans, received a larger share of the degrees conferred despite a decline in the overall number. During the same decade, foreign students also enrolled in large numbers due to decreases in native-born enrollment and institutional financial and research support.[18]

The College during the Period

As the enrollment increased, the University of Arkansas College of

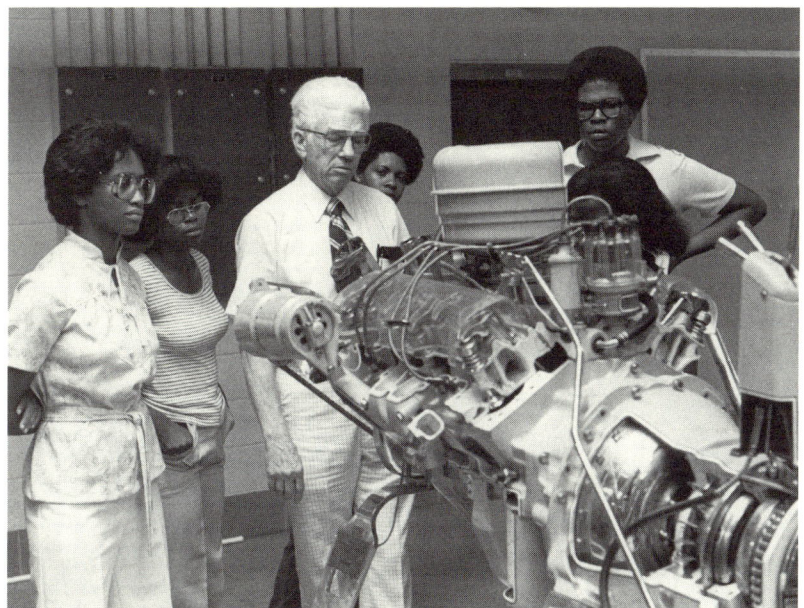

Professor Franklin Deaver and students examine an engine, ca. 1975. (Mullins Library Special Collections, photo collection #4726)

Engineering expanded offerings by improving the Cooperative Education Program and establishing the Engineering Extension Center. The college had used cooperative education since the early 1960s, but the program had never received the proper administrative support. This had changed by 1975 as the college realized the possibilities of expanding enrollment by offering engineering students a way to work in the field between terms and help pay for their education. The Engineering Extension Center, housed in a building across Dickson Street from Engineering Hall, largely conducted research related to research grant projects, but it also provided some courses for graduate students as those already working in the engineering field. The center also relieved some of the pressure from the suddenly crowded conditions of the other engineering buildings. By the fall of 1976, undergraduate enrollment in the College of Engineering totaled 1,059.[19]

The increase in enrollment may have resulted from not only more women and minority engineering students, but also changes in the college due to social and technological issues in the 1970s. In the fall of 1970, the

SAE members. (*Razorback*, 1971)

college offered courses in environmental engineering, and other colleges planned to offer their own classes pertaining to that subject. By the fall of 1976, the degree of bachelor of science in computer science began to be offered.[20]

While celebrations in honor of the university occurred to mark the 1871–1971 milestone, a significant change took place in the College of Engineering as Dean George F. Branigan announced his retirement after serving for twenty-three years. He remarked on his experiences as the third dean of the College of Engineering following William N. Gladson (1913–1936) and George P. Stocker (1936–1948). Branigan recalled accomplishments made during his time such as creating a solid undergraduate program in engineering, which he believed to be the primary mission of the college.[21]

Many beneficial changes occurred in the postwar period during which Branigan served as dean. Full-time faculty increased from forty-one members in 1948 to seventy-eight in 1971, and those qualified by the graduate council to teach graduate-level courses grew from five to seventy-six during the same time period. Only one faculty member held a doctorate in 1948, but forty-six faculty members held doctorates by the spring of

1971. The university awarded over fifty graduate degrees during commencement for the 1969–1970 academic year, whereas none was awarded for the 1948–1949 academic year. As a result of his activities with professional engineering organizations within the state, Branigan succeeded in establishing a requirement that faculty be professional engineers for promotion. He believed students deserved the same level of professionalism that someone in the general public might expect when contracting with an engineer.[22]

The Mechanical Engineering Department might very well owe its existence in a separate building to Branigan. In Engineering Hall, the mechanical engineering lab ran directly under the auditorium where Branigan would often hold meetings. The operation of equipment, especially a large diesel engine, would usually result in a visit by the dean, who would not be pleased by the interruption. As a result, the plans for the new construction in the 1960s called for the Mechanical Engineering Department to be across Dickson Street from the rest of the departments and Dean Branigan's office.[23]

Loren R. Heiple, his successor, took over a successful College of Engineering, but issues of low enrollment figures and employment difficulties for graduates confronted him immediately. However, he felt there had been misleading news accounts of the employment situation, and, in fact, engineers still had opportunities. The problem appeared to be that jobs were not as readily available as in the 1960s.[24]

The University during the Period

By the early 1970s, almost $50 million in construction had been completed on the Fayetteville campus since the 1960s, but even more had already been planned or commenced. In January of 1973, Kimpel Hall, then called the communications building, opened for use, and construction neared completion on the student union, a $6.5 million building. Plans had also been made to renovate Old Main, the old student union, and to construct additions to the school of law and the fine arts buildings. The university also planned new buildings for the business school, the plant sciences department, and the agriculture college, along with renovations and additional facilities for athletics.[25]

The University of Arkansas celebrated its centennial during 1971 with events in Fayetteville, Little Rock, and even Washington, D.C. State government leaders in Little Rock, including Governor Dale Bumpers, held a ceremony in March to begin the year-long celebration. In Washington, D.C., the Arkansas congressional delegation along with more than four hundred friends and alumni in the D.C. area, attended convocation in the National Cathedral, and Senator Fulbright introduced a resolution that expressed congratulations on the part of the House and the Senate, which passed both houses. Fayetteville held its celebration on November 15, 1971, to recognize the official date the town became the choice for the university, and the campus held an event on January 22, 1972, to celebrate the first day of classes. Several other symposiums and publications accompanied the main events over the course of the year.[26]

As the university celebrated the Fayetteville campus centennial, the University of Arkansas system expanded to include campuses in Monticello, Pine Bluff, and Little Rock. Including the three campuses as part of the university system brought the university closer to serving more of the state as its original mission had intended. Each institution remained independent to maintain better control over local issues but would contribute to the university system's goal of adjusting to the shift from a predominantly agricultural state to include industry and technology.[27]

Athletics

The football program continued to have success in the 1970s as the team ranked in the top ten consistently in the late seventies and won two SWC championships. Since the late 1940s, the entire state had been able to follow the Razorbacks' success with the completion of War Memorial Stadium in Little Rock. By the 1970s, the stadium in Fayetteville had received several alterations to increase seating capacity, and the university installed astroturf in 1969. The two venues ensured fans from around the state had a chance to watch the team and establish a connection to the university through the success of the football program. During the fall of 1982, a senior mechanical engineering student, Tom Jones, started as quarterback for the football team. According to Jones, balancing sports and academics helped him manage his time better, and the head coach, Lou Holtz, stressed academics.[28]

Basketball had begun to play just as an important role in fan and alumni support by the 1970s when Eddie Sutton was hired by Frank Broyles as head basketball coach. Under Sutton, the program reached the

NCAA Final Four in 1978 and ended the season in the top ten for three consecutive years. By 1979, the team won 83 games over three seasons, more than any other in the country at the time. By 1978, the cross-country team, under head track coach John McDonnell, won four consecutive SWC championships. The baseball program moved forward when the 1979 team under coach Norm DeBryn reached the College World Series final, and the tennis program finished in the top ten several times in the late 1970s.[29]

Biographical Sketches

Franklin Kennedy Deaver, BSChE (1939), MSME (1960), University of Arkansas; PhD (1969), University of Minnesota, Assistant Professor (1955–1960), Associate Professor (1960–1969), Professor and Department Head (1969–1980), Professor (1980–1984)

Ken Deaver was an Arkansas native whose family was in the lumber business. He served as a lieutenant on submarines in World War II, where his shipmates called him "Danny." Courses he taught included Heat Transfer, Thermodynamics, Machine Design, and Convection. During his time at the university, he worked with many companies including Douglas Aircraft

Franklin K. Deaver, ca. 1982. (Mechanical Engineering collection)

and Phillips Petroleum. He also worked at Sandia Labs. His work resulted in many publications, most dealing with heat transfer. Deaver's high level of scholarship was also evident when he received three National Science Foundation Science Faculty Fellowships during the 1960s. In addition to his time as department head throughout the 1970s, he served as acting department head twice: once in the fall of 1968 and again in the summer of 1980. He and his wife, Gail, lived in Fayetteville after his retirement in 1982. He passed away in 1992.[30]

***Jack H. Cole**, BSME, MSME, PhD (Oklahoma State University, 1968);*
Assistant Professor (1968–1972), Associate Professor (1972–1977),
Professor (1977–1982, 1994–2002)

Jack Cole's industrial experience included work with oil and aircraft companies, and he worked at the North American Rockwell Corporation, Tulsa Division, before coming to Arkansas. Courses that he taught included Fluid Logic, Hydraulic Control, and Dynamics of Machinery, and his teaching focused on machine design and hydraulic and pneumatic control systems. He left the department in 1981 to work for Conoco Oil Company in Ponca City, Oklahoma, in the research and development

Jack H. Cole, ca. 1975. (Mechanical Engineering collection)

department. He returned to the University of Arkansas in 1994. A student in the early 1970s recalled a project he worked on with Cole "to complete the design and build of a special chair for handicapped children. The chair had many adjustable features to accommodate the needs of the child's particular deformity."[31]

Chapter 6

Modern Times, 1980–2004

The Schmidt Years

In 1980, several new faculty members joined the department, but they had already been working for the university. The Engineering Science Department ceased to exist and merged with Mechanical Engineering, which was then named the Mechanical Engineering and Engineering Science Department. Although the new department benefited from the merger of so many talented professors, the first few years were difficult as everyone acclimated to the situation, but many faculty were involved with several research projects in the early 1980s despite the departmental shift.

Department Heads
Mason H. Somerville, 1980–1984
Helmut Wolf, 1984–1985
William F. Schmidt, 1985–present

Faculty
1. Donald A. Gilbrech, BSIE, MSEM, University of Arkansas; PhD, Purdue University; Professor (1980–1990)[1]
2. Ing-Chang Jong, BSCE, National Taiwan University; MSCE, South Dakota School of Mines and Technology; PhD, Northwestern University; Professor (1980–present)
3. Longley R. Kirby, BSCE, University of Tennessee; MSCE, University of Texas; Professor, (1980–1982)
4. Charles W. Crook, BS, BSGE, Oklahoma State University; MS, Brown University; Associate Professor (1980–1995)
5. Mason H. Somerville, BSME, Worchester Polytechnic Institute; MSME, Northeastern University; PhD, Pennsylvania State University; Professor and Department Head (1980–1984)
6. John G. Williams, BA, University of Oxford; PhD, University of London; Visiting Associate Professor (1980–1982), Professor (1983–1987)

7. Gregory R. Gessel, BSChE, MSNucE, PhD, Iowa State University; Assistant Professor (1980–1985)
8. Rick J. Couvillion, BSME, University of Arkansas; MSME, PhD, Georgia Institute of Technology; Assistant Professor (1981–1986), Associate Professor (1986–present)
9. William T. Springer, BSME, MSME, PhD, University of Texas at Arlington; Assistant Professor (1981–1988), Associate Professor (1988–present)
10. Harold A. Sreshta, BSChE, Illinois Institute of Technology; MSMatE, PhD, Drexel University; Assistant Professor (1981–1989)
11. Thomas A. Cook, BSME, Massachusetts Institute of Technology; MSME, University of Arkansas; Instructor (1981–1984)
12. Leon West, BSME, University of Arkansas; PhD, Florida State University; Associate Professor (1982–1990), Professor (1990–present)
13. David A. Renfroe, BSME, Brigham Young University; MSME, PhD, Texas A&M University; Assistant Professor (1982–1988), Associate Professor (1988–1995)
14. Lynn D. Wills, BSME, New Mexico State University; MSME, Northwestern University; PhD, University of Arkansas; Assistant Professor (1984–1990)
15. Charles R. Fagg, BSME, University of Texas; Instructor (1984–1987)
16. William F. Schmidt, BSME, University of Kentucky; MSME, PhD, University of Washington; Professor and Department Head (1985–present)
17. Edward P. Clark, BS, United States Naval Academy; MSAE, University of Michigan; Instructor (1985–1995)
18. Rachel A. Dupuy, BSEM, MSEM, University of Missouri-Rolla; Instructor (1989–1991)
19. Calvin Goforth, BS, University of Texas, Austin; MSME, PhD, Stanford University; Assistant Professor (1990–1995)
20. Carl V. Wikstrom, BA, Rice University; PhD, Massachusetts Institute of Technology; Assistant Professor (1989–1994)
21. Seifollah Nasrazadani, BSME, MSME, PhD, Louisiana State University; Assistant Professor (1990–1995)

22. James D. Landrus, BSME, General Motors Institute; MSME, Purdue University; Instructor (1990–2001)
23. Bobby L. McMasters, BSME, Oklahoma State University; Visiting Instructor (1991–1995)
24. Matthew H. Gordon, BSME, MSME, PhD, Stanford University; Assistant Professor (1992–1997), Associate Professor (1997–present)
25. Mark C. Johnson, BS, Arizona State University; MS, Massachusetts Institute of Technology; PhD, Texas A&M University; Assistant Professor (1992–1994)
26. Larry A. Roe, BSME, MS, University of Mississippi; PhD, University of Florida; Assistant Professor (1993–2000), Associate Professor (2000–present)
27. Michael B. Stewart, BA, Kalamazoo College; MSME, PhD, University of Illinois; Assistant Professor (1993–2000)
28. Glynn P. Adams, BSME, MSME, Louisiana State University; PhD, Purdue University; Assistant Professor (1994–1998)
29. Ajay P. Malshe, BS, S.P. College, University of Poona, India; MS, PhD, University of Poona; Assistant Professor (1994–2000), Associate Professor (2000–present)
30. Saad Alyan, BSME, MSME, PhD, Wichita State University; Visiting Assistant Professor (1994–1995)
31. Arnoldo Muyshondt, BS, MS, Texas Tech University; PhD, Texas A&M University; Assistant Professor (1994–1999)
32. Darin W. Nutter, BSME, MSME, Oklahoma State University; PhD, Texas A&M University; Assistant Professor (1994–2000), Associate Professor (2000–present)
33. James A. Davis Jr., BSME (1989), MSME (1991), PhD (2000), University of Arkansas; Instructor (1995–present)
34. Robert R. Reynolds, BSME, Carnegie-Mellon University; MSME, Purdue University; PhD, Duke University; Assistant Professor (1995–2002)
35. Chao-Hung S. Tung, BSME, National Taiwan University; MSME, PhD, University of Houston; Assistant Professor (2000–present)
36. Stephen A. Batzer, BSME, Michigan Technological University; MSME, GMI Engineering and Management Institute; PhD,

Michigan Technological University; Assistant Professor
(1999–2003)

37. Deepak G. Bhat, BSMetE, University of Poona, India;
MTechMetE, IIT Bombay; PhD, University of Southern
California, Erma Fitch and Raymond Giffels Professor
(2001–present)

38. Min Zou, BSAE, MSAE Northwestern Polytechnical University;
MSME, PhD, Georgia Institute of Technology; Assistant
Professor (2003–present)

In 1980, Mason Somerville became department head, and the department merged with Engineering Science, becoming the Mechanical Engineering and Engineering Science Department. Core courses that all engineering students had taken in the Engineering Science Department

Mason Somerville, ca. 2000. (Mechanical Engineering
collection)

would now be offered through the merged department. University administrators had determined that the Engineering Science Department did not produce enough graduates to justify its existence.[2]

The freshman requirements remained the same for the most part, but upperclassmen requirements changed. Also, the department adjusted the hours so that a senior would only have fourteen hours to complete in the spring. A science elective and mechanical engineering elective lab also changed the senior year. Those course requirements remained the same throughout Somerville's tenure as department head.[3]

Curriculum, 1981

Freshman
Fall, 16 credit hours
Engl, Composition, 3 credits
Chem, College Chemistry, 5 credits
Math, Calculus I, 5 credits
GE, Graphics, 2 credits
GE, Orientation, 1 credit

Spring, 16 credit hours
Engl, Technical Composition, 3 credits
Math, Calculus II, 5 credits
Phys, University Physics I, 3 credits
Phys, University Physics Lab I, 1 credit
ME, Introduction to Mechanical Engineering and Engineering Science, 1 credit
Humanistic-Social Studies Elective, 3 credits[4]

Sophomore
Fall, 17 credit hours
Phys, University Physics II, 3 credits
Phys, University Physics Lab II, 1 credit
Math, Calculus III, 3 credits
ME, Metallurgy and Materials, 3 credits
ME, Statics, 3 credits
Econ, Principles of Economics, 3 credits
IE, Introduction to Computers, 1 credit

Spring, 18 credit hours
Math, Differential Equations, 3 credits
ME, Dynamics, 3 credits
ME, Production Engineering, 3 credits
ME, Thermodynamics I, 3 credits
EE, Electric Circuits I, 3 credits
Humanistic-Social Studies Elective, 3 credits

Junior
Fall, 17 credit hours
ME, Mechanics, 3 credits
ME, Mechanics of Materials, 3 credits
ME, Thermodynamics II, 3 credits
ME, Lab I–Measurements I, 2 credits
EE, Electronics I, 3 credits
Humanistic-Social Studies Elective, 3 credits

Spring, 17 credit hours
ME, Vibrations and Machine Dynamics, 3 credits
ME, Design Stress Analysis, 3 credits
ME, Mechanics of Fluids, 3 credits
ME, Lab II–Measurements II, 2 credits
Technical Elective, 3 credits
Humanistic-Social Studies Elective, 3 credits

Senior
Fall, 17 credit hours
ME, Machine Element Design, 3 credits
ME, Heat Transfer, 3 credits
ME, Lab III–Applied Lab, 2 credits
ME, Elective, 6 credits
Science Elective, 3 credits

Spring, 14 credit hours
ME, Creative Project Design, 3 credits
ME, Elective Lab, 2 credits
ME, Elective, 3 credits

Technical Elective, 3 credits
Humanistic-Social Studies Elective, 3 credits[5]

The Schmidt Years Begin

Mason Somerville oversaw the transition from the Deaver years to the Schmidt years. The anxiety resulting from the merger of two departments and his aggressive style made the Somerville years tumultuous. Somerville departed after the 1983–1984 academic year to become dean of engineering at Texas Tech University, and Helmut Wolf became interim department head for the 1984–1985 academic year. The name of the department changed under Wolf from Mechanical Engineering and Engineering Science to simply Mechanical Engineering. During the summer of 1985, William F. Schmidt became department head, and the Schmidt years began.[6]

Curriculum, 1985

Freshman
Fall, 17 credit hours
Engl, Composition, 3 credits
Chem, General Chemistry, 4 credits
Math, Calculus I, 5 credits
GE, Graphics, 2 credits
Humanistic-Social Studies, 3 credits

Spring, 16 credit hours
Chem, General Chemistry, 4 credits
Math, Calculus II, 5 credits
Phys, University Physics I, 3 credits
Phys, University Physics Lab I, 1 credit
ME, Introduction to Mechanical Engineering, 3 credits

Sophomore
Fall, 16 credit hours
Phys, University Physics II, 3 credits
Phys, University Physics Lab II, 1 credit
Math, Calculus III, 3 credits

ME, Metallurgy and Materials, 3 credits
ME, Statics, 3 credits
Econ, Economics I, 3 credits

Spring, 18 credit hours
Math, Differential Equations, 3 credits
ME, Dynamics, 3 credits
ME, Thermodynamics I, 3 credits
EE, Electric Circuits I, 3 credits
Engl, Technical Composition, 3 credits
Humanistic-Social Studies Elective, 3 credits[7]

Junior
Fall, 17 credit hours
ME, Mechanics of Materials, 3 credits
ME, Thermodynamics II, 3 credits
ME, Lab I–Measurements I, 2 credits
EE, Electronics I, 3 credits
ME, Mechanics of Fluids, 3 credits
Humanistic-Social Studies Elective, 3 credits

Spring, 17 credit hours
ME, Design Stress Analysis, 3 credits
ME, Lab II–Measurements II, 2 credits
ME, Heat Transfer, 3 credits
ME, Mechanisms, 3 credits
Math Elective, 3 credits
Humanistic-Social Studies Elective, 3 credits

Senior
Fall, 17 credit hours
ME, Lab III–Applied Lab, 2 credits
ME, Production Engineering, 3 credits
ME, Vibrations and Machine Dynamics, 3 credits
ME, Thermal Systems Analysis and Design, or
ME, Machine Element Design, 3 credits
ME, Elective, 6 credits

Spring, 14 credit hours
ME, Creative Project, Mechanical, or
ME, Creative Project, Thermal, 3 credits
ME, Elective Lab, 2 credits
ME, Elective, 3 credits
Technical Elective, 3 credits
Humanistic-Social Studies Elective, 3 credits[8]

Despite the efforts to attract highly qualified students, the department had experienced retention problems by the mid-1990s. The crux of the issue seemed to be that students found themselves alone too often with few classmates from previous classes. The department worked with the assistant dean of engineering to solve the problem by grouping freshman students together in blocks of classes so camaraderie and support would develop more often. Computerized registration systems made this strategy difficult to implement, but the department attempted to accomplish something similar by grouping freshmen together during orientation classes.[9]

The department restructured the curriculum for the 1994–1995 academic year. A new requirement for the freshmen included a completely new course, MEEG 1113 that would have a laboratory component. The new course allowed for more actual experimental work in the first year. The senior year design course increased to five from three hours by reducing other requirements and would be taken over two semesters. For the first semester, MEEG 4132, students chose the projects, and then during the next semester in MEEG 4133, students worked on their projects.[10]

By 1997, the Mechanical Engineering Department had existed for one hundred years, and requirements differed considerably from those in the late nineteenth century.

Curriculum, 1997

Freshman
Fall, 16 credit hours
Engl, Composition I, 3 credits
Chem, University Chemistry I, 3 credits
Chem, University Chemistry Lab I, 1 credit

Math, Calculus I, 4 credits
GE, Engineering Graphics, 2 credits
ME, Intro to Mechanical Engineering I, 3 credits

Spring, 18 credit hours
Chem, University Chemistry II, 3 credits
Chem, University Chemistry Lab II, 1 credit
Math, Calculus II, 4 credits
Phys, University Physics I, 3 credits
Phys, University Physics Lab I, 1 credit
Engl, Technical Composition II, 3 credits
ME, Intro to Mechanical Engineering II, 3 credits

Sophomore
Fall, 17 credit hours
Phys, University Physics II, 3 credits
Phys, University Physics Lab II, 1 credit
Math, Calculus III, 4 credits
ME, Introduction to Materials, 3 credits
Econ, Basic Economics, 3 credits
ME, Statics, 3 credits

Spring, 16 credit hours
Math, Differential Equations, 4 credits
ME, Dynamics, 3 credits
ME, Thermodynamics, 3 credits
EE, Electric Circuits and Machines, 3 credits
Humanities-Social Sciences Elective, 3 credits

Junior
Fall, 17 credit hours
ME, Mechanics of Materials, 3 credits
ME, Thermodynamics II, 3 credits
ME, Mechanical Engineering Lab I, 2 credits
EE, Engineering Electronics, 3 credits
ME, Mechanisms, 3 credits
ME, Numerical Methods I, 3 credits

Spring, 17 credit hours
ME, Vibrations and Machine Dynamics, 3 credits
ME, Design Stress Analysis, 3 credits
ME, Mechanical Engineering Lab II, 2 credits
ME, Mechanics of Fluids, 3 credits
ME, Elective, 3 credits
Humanities-Social Sciences Elective, 3 credits

Senior
Fall, 13 or 16 credit hours
ME, Heat Transfer, 3 credits
ME, Machine Element Design, 3 credits
ME, Creative Project Design I, 2 credits
ME, Mechanical Engineering Lab III, 2 credits
ME, Design Elective, 3 credits
Humanities-Social Sciences Elective, 3 credits

Spring, 15 or 18 credit hours
ME, Elective, 3 credits
ME, Thermal Systems Analysis and Design, 3 credits
ME, Creative Project Design II, 3 credits
Technical Elective, 3 credits
Humanities-Social Sciences Elective, 6 credits[11]

The early 1980s proved to be a particularly active time for the mechanical engineering faculty. Professor Cecil Cogburn received assistance from Professors Leon West and John Williams on nuclear dosimetry projects, an example of the research begin done at the SEFOR calibration center. Mason Somerville began working with Cogburn and West on establishing a full-range regional radiation calibration laboratory with the National Bureau of Standards and with the help of U.S. senator Dale Bumpers. Professor Henry Hicks worked on a project studying conversion of waste to charcoal. Professors William Springer and Rick Couvillion worked on a two-year program funded by the Veterans' Administration Hospital in Little Rock to develop analytical and laboratory models of the bladder and urinary tract. Professor Helmut Wolf directed a project on the analytical investigation of energy movement and the associated stress caused by thermal expansion in materials used by nuclear power plants. Another project involving Professors Harold Sreshta, Rick Couvillion, David Renfroe, and Mason Somerville examined ways to control, reduce, and possibly eliminate ash slagging and fouling at the Southwestern Power Company Flint Creek, Arkansas, power plant.[12]

Dickson Street looking west toward the campus, 1981. (Mullins Library Special Collections, Kent R. Brown Collection, Miller #19)

By the end of the 1980s, finding replacements for retiring senior faculty members proved to be challenging, as a large pool of qualified applicants did not exist. However, the lack of candidates suggested that students with a PhD in mechanical engineering would have opportunities in the education field. The department succeeded in finding two professors when Mark Johnson and Matthew Gordon joined the faculty in 1991. Once the department began hiring more professors, a sequence of required courses began in the fall of 1993 to facilitate new faculty research. The change resulted in required courses offered less frequently than every semester, which the department anticipated would give faculty more time to research and develop a career, ultimately providing graduate students with more opportunities.[13]

The Mechanical Engineering Department had been moving in the research direction for many years by then. Beginning in the 1960s, research began to be emphasized as more money became available to work on projects such as aerospace technology. Although faculty had always participated in research before the 1960s, the low participation resulted in small praise for projects. However, as the department hired more faculty members with PhDs and as funding from the National Science Foundation and other sources increased throughout the 1970s and 1980s, the number of courses a professor was expected to teach by the 1990s had decreased to allow more time for research.[14]

The number of department faculty members continued to fluctuate in the mid-1990s. While Darin Nutter joined the department in August of 1994 and Robert R. Reynolds in January of 1995, E. Paul Clark retired in December of 1994. Three other faculty members resigned to work in private sector, but Jack Cole returned to the department after a career in industry for twelve years. This movement between the public and private sector also occurred across the country as industries and universities began to cooperate more to develop technology more expeditiously. The competition between the United States and Japan in the mid-1980s fostered much of this technological drive, as well as grants from federal government agencies such as the National Science Foundation.[15]

In the early 1980s, mechanical engineering students frequently participated in an event that directly applied what they had learned in class. During the 1980 fall semester, students in the senior design class began work on a mini-baja car for the 1981 mini-baja competition in Cookeville, Tennessee, sponsored by the ASME. Several criteria had to be followed to make the vehicle have consumer appeal, meet safety standards, be a creative design, and pass an endurance and maneuverability course. Also, the car could not cost more than $1,200. Students had spent most of their time working on the frame two nights a week and on some weekends. Although the group could not raise a significant amount of funds from local businesses, the students received donations during Engine Week.[16]

The students who participated in 1981 laid a good foundation for the 1982 baja team (Professor Kimzey taught the fall 1981 design class and served as advisor for the project), which raced to a third-place finish at the competition in Rochester, New York. The amount teams could spend had increased to $1,350, and several industries donated parts. Briggs and Stratton donated the engines for all of the entries nationwide, and other parts were donated by Arkansas industries. Along with the University of Arkansas chapter of the Society of Automotive Engineers, the ASME entered several vehicles in the mini-baja competition to be held at the University of West Virginia in the spring of 1984. The University of Maryland had hosted the competition the previous spring, and the vehicles from the U of A's team placed fourteenth and fifteenth in a field of sixty. In 1992, after a seven-year absence, mechanical engineering students in SAE entered the competition and won first place in design originality and third place in the hill-climbing contest. The group also participated the next year in El Paso, Texas.[17]

The ASME began the fall of 1983 with a picnic at Somerville's house. The group planned several activities for the semester that included a locker rental program and a computerized book exchange to be available to all university students. The idea for the computerized book exchange may have been due to the introduction of a computerized message service that fall.[18]

In the spring of 1986, the organization received the engineering college activity award for 1985–1986 and joined students from Lamar

ASME members. (*Razorback*, 1980)

University on a field trip to NASA in Houston. The high level of chapter participation did not go unnoticed by the regional ASME. In the fall of 1986, students found out that the chapter had been placed second among thirty-three schools in the region based on activities and involvement.[19]

The Mechanical Engineering Department began coordinating a program called A World in Motion in 1989. The SAE, along with faculty advisor David Renfroe, began the program as part of Vision 2000, and the university became the only higher education institution involved with A World in Motion. Engineering students visited six Fayetteville elementary schools to encourage interest in science during a time some children began to lose interest in the subject.[20]

Although the university had many problems with funding faculty and staff positions, student enrollment continued to rise. Freshman enrollment increased by 20 percent for the 1985–1986 academic year as well as the quality, which improved by 5 percent based upon ACT scores. The improved quality of the freshman class helped to strengthen the mechanical engineering program, which was already considered strong as a result of the current students' high achievements. In order to continue the development of highly qualified graduates, the Mechanical Engineering Department developed several new courses in theory of aeronautics, advanced power systems, jet propulsion, and air conditioning. The department also revised laboratory and design courses. One of those, the senior design course, was adjusted by the department to address safety and ethics topics that would be subject to change by anticipated accreditation requirements. The Mechanical Engineering Department held a one-week workshop for faculty on computers with good results as ten of them went on to revise courses to include more computer work.[21]

A mechanical engineering student from the early 1960s reflected on the increase of technological advances, fewer required lab courses, and the possibility that "students have only 'simulation' experiences that lack any feel for underlying processes and real system behaviors. I almost wish we could return to slide rules so students today would not think that there were a gazillion significant figures in every answer. And while I appreciate the increased precision of modern computational methods, there is something lost in the art of modeling with analytical solutions that provided a feel for the dominant trend that is totally lacking in the computational tools."[22]

Even though the economy struggled in the mid-1980s, employment opportunities remained available to mechanical engineering graduates. Many of the largest companies in the United States required employees with BSME degrees to design equipment, especially those companies involved in the manufacturing of electronic components. In the research areas of these same companies, advanced development of future manufacturing designs provided employment for those students with PhDs, who also found employment at universities as well. The growth in the electronics industry and other large companies continued to provide jobs for mechanical engineers into the late 1980s as nearly all of the graduates found employment. In fact, so many jobs existed that many went unfilled nationally and in Arkansas.[23]

According to information in the spring of 1981, mechanical engineering graduates averaged $22,680 in annual starting salaries, with the highest being $25,800 and the lowest at $18,696. By the fall of 1984, average annual salaries for mechanical engineering graduates had increased by several thousand dollars to $26,280. Only two other fields averaged more but not by very much. Electrical engineering graduates expected $26,556, and chemical engineering took the lead with $27,420. Those with an MSME could also expect salaries to fall below electrical and chemical engineering as well, but not by much. A mechanical engineering graduate with a master's averaged $30,288, a chemical engineering degree averaged $30,684, and an electrical degree averaged $31,008. Salaries for mechanical engineering graduates receiving diplomas in 1990 averaged $34,000 with a BSME and $37,000 with a MSME. Utility, oil, and chemical companies usually constituted the bulk of potential employers.[24]

A year before construction commenced on the Bell Engineering Center in 1984, the Mechanical Engineering Department proceeded with plans to enclose the front plaza of its building. Professor Couvillion recalled how the project almost failed to be completed after "Department head Mason Somerville had raised $14,000 for construction. This was far, far short of the amount required to create the great student space we have now. However, Gene Billingslea, one of our Academy members, was a deputy director at the Physical Plant and supervised the construction. The work was completed with the money available. We didn't ask questions. Our largest classroom, ME 212, was modernized recently and is named in his memory."[25]

The front porch of the Mechanical Engineering Building was enclosed in 1983 to create a student lounge, 2003. (Mechanical Engineering collection)

The Mechanical Engineering Department continued to attract qualified high school graduates as the average composite ACT remained greater than 26 by 1990, and the department made every effort to raise this statistic. The Arkansas Academy of Mechanical Engineers (AAME), an organization of distinguished graduates from the program, offered scholarships to students with a score of 30 or above on the ACT, and the department personally contacted these same students. The AAME had originally organized in 1982 when Professors Ken Deaver, Donald Gilbrech, and Cecil Cogburn developed a list of prominent engineering graduates to become the charter members. Once established, the organization began collecting funds to assist in mechanical engineering education in ways the state did not. In addition to raising scholarship money, the AAME also provided funds to renovate a classroom in the Mechanical Engineering Building subsequently renamed the Arkansas Academy of Mechanical Engineering Lecture Hall.[26]

Mechanical Engineering Building, 2003. (Mechanical Engineering collection)

Significant Curriculum Change

The last significant curriculum change occurred at the beginning of the Deaver years in 1969, when the curriculum was reduced from 141 to 132 hours. In the years that followed, some minor adjustments were made every few years. After consideration of the department's objectives and examination of curricula at a number of schools, the faculty made a major change for the fall 2002 semester. The number of hours was reduced to 124, giving students a realistic chance to graduate in four years. Students were also given much more flexibility in their electives, allowing them to pursue a pre-med curriculum or earn minors in business or one of the sciences without adding many additional hours.

Curriculum, 2002

Freshman
Fall, 16 credit hours
Engl, Composition I, 3 credits
Chem, University Chemistry I, 3 credits

Chem, University Chemistry Lab I, 1 credit
Math, Calculus I, 4 credits
GE, Engineering Graphics, 2 credits
ME, Intro to Mechanical Engineering I, 3 credits

Spring, 15 credit hours
Chem, University Chemistry II, 3 credits
Chem, University Chemistry Lab II, 1 credit
Math, Calculus II, 4 credits
Phys, University Physics I, 3 credits
Phys, University Physics Lab I, 1 credit
Engl, Technical Composition II, 3 credits

Sophomore
Fall, 14 credit hours
Phys, University Physics II, 3 credits
Phys, University Physics Lab II, 1 credit
Math, Calculus III, 4 credits
ME, Introduction to Materials, 3 credits
ME, Statics, 3 credits

Spring, 16 credit hours
Math, Differential Equations, 4 credits
ME, Dynamics, 3 credits
ME, Thermodynamics, 3 credits
EE, Electric Circuits and Machines, 3 credits
ME, Computer Methods, 3 credits

Junior
Fall, 17 credit hours
ME, Mechanics of Fluids, 3 credits
ME, Mechanics of Materials, 3 credits
ME, Mechanical Engineering Lab I, 2 credits
ME, Machine Dynamics and Control, 3 credits
EE, Engineering Electronics, 3 credits
Humanities-Social Sciences Elective, 3 credits

Spring, 14 credit hours
ME, Heat Transfer, 3 credits
ME, Mechanical Engineering Lab II, 2 credits
Tech-Science Elective, 3 credits
Economics, 3 credits
Humanities-Social Sciences Elective, 3 credits

Senior
Fall, 17 credit hours
ME, Thermal Systems Analysis and Design, 3 credits
ME, Machine Element Design, 3 credits
ME, Creative Project Design I, 3 credits
ME, Mechanical Engineering Lab III, 2 credits
Tech-Science Elective, 3 credits
Humanities-Social Sciences Elective, 3 credits

Spring, 15 credit hours
ME, Creative Project Design II, 3 credits
Tech-Science Elective, 6 credits
Humanities-Social Sciences Elective, 6 credits

The College during the Period

After struggling for several years with an inadequate staff, the *Arkansas Engineer* ceased publication after the spring 1985 issue but returned in the summer of 1986. Tau Beta Pi members made up the staff when the periodical resumed publication because the honor society had enough members to handle the responsibility. Engineer's Day had also fallen by the wayside for a period of time in the mid-1980s. After a two-year absence, Engine Week returned in the spring of 1988 on April 9, but would remain much more subdued than celebrations in the past.[27]

As a result of the poor economic climate in the state and the country in the early 1980s, the university received much less in appropriations from the legislature. The decline in funding from the state directly impacted facilities and faculty for instruction and research. The financial situation had not improved much by the mid-1980s as the University of Arkansas struggled to provide enough faculty and staff to work in and maintain the programs and buildings on the campus. Even though problems with operations existed, the North Central Association of Colleges and Schools visited the campus in 1986 and reported a strong faculty increasingly active in research.[28]

Other departments in the college found themselves redesigning courses as well to adjust to trends that had been developing for several decades. After World War II, engineering education slowly shifted from instruction of specific engineering skills to ensuring that all engineering students graduated with a certain level of skills in fundamental areas. By the 1980s, computer literacy was one of those fundamental skills that students needed to have successful careers, and the college integrated computers into the curriculum to achieve accreditation. Courses had to be redesigned in order to have a certain level of computer operations in the classroom and the lab so students would be exposed to computers as much as possible.[29]

The college made several strides toward expanding its classroom space by the early 1980s as plans moved forward with the 142,500-square-foot Bell Engineering Center on campus, the 200,000-square-foot Engineering Experiment Station in the former Bear Brand Hosiery building south of campus, and the renovation of the existing 63,750 square feet in Engineering Hall. The rate of growth in the college led to the planning of a new engineering complex in the fall of 1979, and almost a year later, the requirements and other specifications of the building had been planned sufficiently to proceed. Obviously, the building had to have more laboratory, classroom, and office space. Other considerations included a large auditorium to avoid scheduling classes in other parts of the campus away from the engineering departments and placement of the structure so it would not obscure the view of Old Main. The area on South Campus Drive provided the space for the 142,500-square-foot building as well as placing it between already existing engineering structures. Concerns over continued accreditation of the engineering programs by the Accreditation Board of Engineering and Technology also encouraged the college to plan for the new building, as well as acquire the new facility for the Engineering Experiment Station.[30]

Construction began in May of 1984, and the college officially began to use the new Bell Engineering Center in January of 1987. The decision of the name came after Melvyn E. Bell, BSEE 1960, pledged $8 million to the university's Engineering Excellence Campaign. Both Engineering

Hall and the Mechanical Engineering Building received updated computer network wiring in order to link all of the engineering facilities together. The new equipment that accompanied this infrastructure update included two super minicomputers, eight minicomputers, and more than two hundred microcomputers. With the renovation of the Engineering Hall, the college occupied three buildings instead of being spread out across many buildings as it had been for years. Agricultural Engineering and Computer Science Engineering, along with Research Services, moved into Engineering Hall, which received attention as money slowly became available over the next three years.[31]

By the beginning of the twenty-first century, the Engineering Research Center, often called Engineering South, reflected a trend in collaboration among many of the engineering departments. Whereas research work had been largely completed by separate departments in the past, the move toward more cooperation has produced larger funds for research, especially in the fields of nanotechnology, material science, and biological research. There have even been projects with other universities in the region, such as astronautics work with Oklahoma State University.[32]

Closing

During fall 2003, William F. Schmidt announced that he would end a nineteen-year tenure as department head, effective July 2004, but would remain as a mechanical engineering faculty member. New engineering dean Ashok Saxena began the process of searching for his replacement. The college and mechanical engineering are beginning a new era.

This new era begins with a number of growing research programs. The Mechanical Engineering Department has programs in materials and manufacturing, micro/nanotechnology, space/planetary sciences, and energy systems in the million-dollar range. These programs will greatly expand the graduate student population and produce more PhD graduates who will start high-technology companies in the state.

With strong support from engineers across the state, the mechanical engineering program is set to accomplish even more great things over the next one hundred years and beyond. The technical aspects of the program at the start of the new millennium would have been impossible for the early professors to have envisioned, but they established a solid foundation on which to build. Throughout the years, the department always

Mechanical Engineering faculty and staff, Fall 2003. *Front Row:* Judy Goodner-Holt (staff), Susan Ray (staff), Larry Roe, Linda Pate (staff), Darin Nutter, Ben Fleming (machinist), Deepak Bhat. *Back Row:* Leon West, Larry Brown (technician), Rick Couvillion, Ajay Malshe, Matt Gordon, William Springer, William Schmidt (department head), Steve Tung, Ing-Chang Jong. Not pictured: Min Zou, James Davis. (Mechanical Engineering collection)

maintained a balance between faculty and facilities, never allowing one to surpass the other. The work had definitely paid off by the new millennium as students continued to achieve new heights.

Biographical Sketches

William F. Schmidt, BSME, University of Kentucky; MSME, PhD, University of Washington; Professor and Department Head (1985–present)

William F. Schmidt became the department head of mechanical engineering at the University of Arkansas in 1985. Prior to coming to the

William F. Schmidt, 2003. (Mechanical Engineering collection)

University of Arkansas he was chair of the Mechanical Engineering Department and director of the Computer Aided Design Lab at the University of Maine in Orono, where he began his academic career in 1968 as an assistant professor. While at the University of Maine he taught courses in mechanics, finite elements, and design, and conducted research in the area of nonlinear finite element analysis. In 1980, he was selected to be chair of the department at Maine.

He began his college education at the University of Kentucky, earning a BS degree in mechanical engineering in 1964. During his junior and senior years of college he worked as a product test engineer for IBM in their office products division in Lexington, Kentucky. Upon graduation, he accepted employment with Exxon at the Engineering Research Center in Florham Park, New Jersey. The work was in the machinery division and focused on vibration issues in rotating equipment. It was during this period of time that the desire to become a college professor came to the forefront and a decision was made to leave Exxon and return to college. He was awarded a National Science Foundation graduate fellowship to study at the University of Washington in Seattle, Washington, where the MS and PhD degrees were obtained. Concurrent with the final stages of work on the PhD, he was employed by Mathematical Sciences Northwest, a consulting firm that did work in the aerospace industry. His primary responsibility was the development and use of finite element programs for analyzing solid propellant rocket engines.

Dr. Schmidt has been active in his service to the university and the college. He is a member of a number of university committees including the High Density Electronics Center Steering Committee that coordinates the interdisciplinary work of the center on the Fayetteville campus. He also is involved nationally in the ASME and has served as secretary, vice chair, and chair of the ASME Region I department heads group while at Maine, and the Region X group after locating to Arkansas. In ASME he has had and continues to have a number of national committee activities including being a member of the board for engineering education, chair of the Graduate Fellowship Committee, and a member of the nominating committee. He has served as a board of director for ASEE and also was appointed to membership on the executive committee for the Mechanical Engineering Division of ASEE. He has served as chair of the Ralph Coats Roe Award Committee (a $10,000 award given to a mechanical engineering faculty member), as secretary to the Professional Engineers in Education board for NSPE, and as a board of director for NSPE. He currently is a licensed engineer in Arkansas and serves on the ASPE board.

His research has resulted in fifty-six referred journal publications, thirty-one presentations, and one book chapter. The current research is in the general area of nondestructive evaluation, electronic packaging, and MEMS with a focus in the area of mechanical design issues. In 2001 he was awarded the ASME Dedicated Service Award. He has also been recognized for his teaching and in 1983 he was awarded the Western Electric Fund Award for Excellence in Teaching.

William T. Springer, *BSME, MSME, PhD, University of Texas at Arlington; Assistant Professor (1981–1988), Associate Professor (1988–present)*
Dr. William Springer is a registered professional engineer in Arkansas. During his tenure with the department he has taught both undergraduate

William T. Springer, ca. 1981. (Mechanical Engineering collection)

and graduate courses that include Mechanical Engineering Fundamentals, Dynamics, Mechanics of Materials, Mechanisms, Vibrations, Design Stress Analysis, Machine Element Design, Advanced Machine Design, Structural Dynamics, and Modal Analysis.

Dr. Springer's research interests include experimental modal analysis, structural dynamics, structural integrity monitoring, nondestructive evaluation, biomechanics, vibration testing, and holographic interferometry applications. He has published thirty-one papers, contributed to the *Handbook on Structural Testing*, made several invited presentations, and written several reports on industrial projects. In addition, he has attended several short courses, participated in developing a distance learning program at the U of A, reviewed papers for several archival journals, and served as the technical editor for the *International Journal of Experimental and Analytical Modal Analysis*.

Dr. Springer has received the Honorary Fellow Award, Southwest Chapter of the American Helicopter Society; the Carl W. Files Service

Award from the University of Texas at Arlington mechanical engineering department; the Outstanding Research Award for the Mechanical Engineering Department (twice); the Chancellor Teaching-Mentor Summer Support Award; the College of Engineering Teaching Innovation Award (twice); and the University Teaching Academy Teaching Enhancement Award. He was also a NASA/ASEE Summer Faculty Fellow at Marshall Space Flight Center, Alabama, in 1987 and 1988.

Dr. Springer is a member of ASME, SAE, SEM, and Tau Beta Pi, where he served on the executive committee and as chairman of the NDE Engineering Division of ASME, and the Modal Analysis and Dynamic Systems Division of SEM. He has also organized numerous technical sessions and assisted in organizing technical meetings for both ASME and SEM.

Rick J. Couvillion, BSME (1975), University of Arkansas; MSME (1978), PhD (1981), Georgia Institute of Technology; Assistant Professor (1981–1986), Associate Professor (1986–present)
Before beginning his career at the University of Arkansas, Rick Couvillion was a presidential fellow in the Georgia Tech School of Mechanical Engineering and Plant Project Mechanical Engineer at the Aluminum Company of America in Bauxite, Arkansas. He became a professional engineer in his first semester at the U of A. He has taught courses in the undergraduate and graduate levels in the areas of thermal sciences, computer methods, and electronics packaging. His work with the curriculum resulted in the transformation of the course MEEG 3703 from Numerical Methods to Computer Methods in Mechanical Engineering and the development of MEEG 4483—Thermal Systems Analysis and Design. He has been named the outstanding teacher in the department five times and outstanding researcher three times. In 1984–1985 he won both awards in the same year, something no one else has done in the history of the departmental awards. He was also recognized as an outstanding educator by the Arkansas Academy of Mechanical Engineers, Region X of ASME, and the Society of Automotive Engineers.

Couvillion is a member of Tau Beta Pi and Pi Tau Sigma, for which he serves as the faculty advisor for the U of A chapter. He served many years as faculty advisor of the ASME student section and is currently the faculty advisor for the ASHRAE student chapter. He also now serves as

Rick J. Couvillion, ca. 1981. (Mechanical Engineering collection)

the ASME Region X assistant vice president for education. As an ASHRAE member, he has chaired two technical committees and been editor of a number of ASHRAE handbook chapters.[33]

He and coauthor Dave Hart wrote one of the early books on the analysis and design of earth-coupled heat pump systems.[34]

Leon West, BSME (1968), University of Arkansas; PhD (1973), Florida State University; Associate Professor (1982–1990), Professor (1990–present)

Before coming to the University of Arkansas, Leon West gained field experience through ten years of work in national and industrial laboratories, in addition to receiving a National Science Foundation grant during his time at Florida State University. Much of his work in the Mechanical Engineering Department was in the field of nuclear research. He was the founder and director of the Southwest Radiation Calibration Center, which grew out of the work at the SEFOR, and he served as officer in multiple American Nuclear Society divisions, which awarded him an out-

Leon West, ca. 1982. (Mechanical Engineering collection)

standing service award. He was also the recipient of a research award from the University of Arkansas College of Engineering.[35]

According to West, "I became (and remain) a professor at the University of Arkansas because of the strong commitment I feel to the young people of Arkansas who seek better educational opportunities." To this effect, he was awarded the University of Arkansas Mortar Board Outstanding Professor Award and a teaching award from the college. His dedication to students was also evident in the many years of service as the SAE chapter advisor.[36]

Matthew H. Gordon, BSME, MSME, PhD, Stanford University;
Assistant Professor (1992–1997), Associate Professor (1997–present)
Before Dr. Matthew Gordon joined the U of A Mechanical Engineering Department, he was a graduate student at Stanford University where he completed both experimental and theoretical evaluations of thermo-dynamic and transport plasma properties in a nonequilibrium environment. Since 1992, Matt has continued his study of nonequilibrium

Matthew H. Gordon, 2003. (Mechanical Engineering collection)

plasmas at the University of Arkansas through both numerical and experimental investigations of microwave and magnetron plasma systems used to grow various films. He has extensively studied thermal-management and thermal-stress issues in high-performance electronic packages involving such novel materials as diamond and carbon nanotubes. Additionally, he has studied highly nonequilibrium material processing systems involving lasers as the thermal source. To date, his work has led to the graduation of nearly twenty MS and PhD candidates, numerous published papers, and millions of dollars in outside research support. He has chaired sessions at professional meetings, has reviewed several papers and proposals, and has regularly spoken on these topics at national and international meetings. Complementing his research, Matt is the faculty advisor for the local ASME student section, and regularly teaches courses in thermodynamics, heat transfer, fluid dynamics, numerical methods, experimental laboratory methods, and finite elements. In addition to earning the Fred M. Carter Award for obtaining the highest score on the professional engineers examination in Arkansas during 1995, Dr. Gordon was also recognized as the department's outstanding researcher in 1996.

Larry A. Roe, BSME (1971), MS (1976), University of Mississippi; PhD (1987), University of Florida; Assistant Professor (1993–2000), Associate Professor (2000–present)

Larry Roe worked for Pratt-Whitney Aircraft and Virginia Tech University before joining the faculty in the fall of 1994 as an assistant professor. Since coming to the department, he has developed six new 4000–5000 level courses. In addition to work with the curriculum, he has been involved with jet propulsion research in university labs and

Larry H. Roe, 2003. (Mechanical Engineering collection)

elsewhere. He was instrumental in the development of the combustion/propulsion lab at the University of Arkansas Engineering Research Center, and he was selected as a NASA Summer Faculty Fellow in 1998 and 2000. In cooperation with Oklahoma State University, he helped create the Arkansas-Oklahoma Center for Space and Planetary Studies. In 1995, he won the Fred M. Carter Award for obtaining the highest score on the Professional Engineer's Exam in Arkansas.[37]

Darin W. Nutter, BSME, MSME, Oklahoma State University; PhD,
 Texas A&M University; Assistant Professor (1994–2000), Associate
 Professor (2000–present)
Prior to attending Texas A&M University and getting his PhD in their
Energy Systems Lab, Darin Nutter worked as a project engineer with the
design and construction of refrigerated warehouses for Fleming Foods,
Inc., of Oklahoma City. He has been a registered professional engineer
since 1993.

Darin W. Nutter, 2003. (Mechanical Engineering collection)

He has taught several mechanical engineering courses including
Introduction to Mechanical Engineering, Thermodynamics II, Thermal
Systems Laboratory, Industrial Waste and Energy Management, and
Indoor Environmental Design. He has received four departmental teach-
ing awards.

Nutter's research interests are encompassed within the broad area of
energy systems. Including both experimental and numerical studies, Dr.
Nutter's background and recent focus has been on the improvement of heat-

ing, ventilating, air-conditioning, and refrigeration (HVAC&R) systems with regard to operation and energy efficiency. Fundamental areas of interest include heat transfer, thermodynamics, and liquid-vapor phase change. He has significant experience and interest in the applied areas of energy consumption, utilities, and energy conservation related to industrial, commercial, and residential buildings and/or systems. To date, he has received $750,000 of outside research funding, leading to nearly fifty publications including refereed journal and conference papers and technical reports.

Nutter has been active on both university committees and national engineering societies. For example, he has served two terms on the Faculty Senate and Academic Standards Committee. Dr. Nutter is also the past chair of the American Society of Heating, Refrigerating, and Air-conditioning Engineers (ASHRAE) technical committee (TC) 7.6 (Unitary and Room Air Conditioners and Heat Pumps) and a member of TC 1.3 (Heat Transfer and Fluid Flow) and TC 6.8 (Geothermal Energy Utilization). In addition, he is a member of the American Society of Mechanical Engineers (ASME), the American Society for Engineering Education (ASEE), the International Ground Source Heat Pump Association (IGSHPA), and the American Society for Testing and Materials (ASTM).

Ajay P. Malshe, BS, S.P. College, University of Poona, India; MS, PhD, University of Poona; Assistant Professor (1994–2000), Associate Professor (2000–present)
Dr. Ajay Malshe is also an adjunct-faculty of the Electrical Engineering Department. He is director of the Science and Engineering Research Center (SERC) for Nano and Micro Systems, a multistate research center, and Materials and Manufacturing Research Laboratories.

Malshe is a materials scientist and engineer who has performed multidisciplinary research and education programs in nanomanufacturing, surface engineering for advanced machining and MEMS and microelectronic packaging and integration. He has initiated the development of wafer-level and chip-scale packaging of MEMS and related microsystems, nanoparticle composite coatings, lasers for chemically clean nano-machining, and nano-mechanical machining tools and system-on-a-chip.

He has authored over one hundred refereed publications and two book chapters, and he holds six patents. Dr. Malshe has graduated over twenty graduate students, trained numerous postdoctoral fellows, and provided research experience to several undergraduate and high school students. He has received thirteen awards for research, education, and service achievements (1996–2003) and is listed in Lexinton's *Who's Who*. He serves the profession through various related academic societies and journals in the role of conference chair, session chair, reviewer, editorial board member, and executive council member and committee chair. He has received outside funding of more than $6 million to support his research.

Ajay P. Malshe, 2002. (Mechanical Engineering collection)

He has an extensive record of collaboration globally with academic institutions and companies and has cofounded two hi-tech companies in the state of Arkansas. His family includes wife, Savita, his son, Harsha, and his daughter, Ashvini. Traveling for sightseeing and painting and photography are his hobbies.

James A. Davis Jr., BSME (1989), MSME (1991), PhD (2000), *University of Arkansas; Instructor (1995–present)*

Dr. James Davis's primary responsibilities include teaching and assisting the department head with administrative activities. Prior to his arrival in Fayetteville, Davis had been a project engineer for Eastman Chemical Company. He earned his BSME, MSME, and PhD from the University of Arkansas.

Over the years, Davis has taught courses in the thermal science area. Among the courses he has taught have been Thermodynamics, Mechanics of Fluids, Laboratory I, Laboratory III, and Thermal Systems Analysis and Design. He has served on the committees of three graduate students and supervised nine undergraduate special projects.

Davis has authored seven publications. He has been a registered professional engineer since 1996 and is a member of ASME.

James A. Davis Jr., 2003. (Mechanical Engineering collection)

Chao-Hung S. (Steve) Tung, *BSME, National Taiwan University;*
MSME, PhD, University of Houston; Assistant Professor
(2000–present)

Prior to joining the U of A, Dr. Chao-Hung Tung worked as a Postdoctoral Fellow at the Department of Mechanical and Aerospace Engineering at UCLA from 1993 to 1997. While at UCLA, Steve participated in a joint UCLA/Caltech research group in the development of a ground-breaking MEMS (micro electro-mechanical systems) drag control system. The group successfully fabricated and tested the first functional micro-thermal shear-stress sensor. From 1997 to 1999, he was an engineering specialist for Litton Guidance and Control Systems. At Litton, he was involved in the development of high-precision MEMS inertial sensors, including the first all-silicon micro gyro designed for navigational applications.

At the University of Arkansas, Steve teaches both undergraduate- and graduate-level classes. The undergraduate classes include Dynamics, Fluid Mechanics, and Introduction to Materials. The graduate classes include advanced fluid mechanics, introduction to MEMS, and advanced MEMS. The MEMS classes are new mechanical engineering classes initiated by him when he joined the department in 2000.

His research interest is in the application of advanced micro- and nano-scale technologies toward the fabrication of novel engineering systems and devices. Current research includes the development of a hybrid microfluidic system fabricated by combining artificial devices with nano-biological motors. He is also investigating new polymer/nanoparticle composites in which the behavior of the particles is controlled by MEMS devices. His research is supported by various funding agencies including the National Science Foundation, the Jet Propulsion Laboratory, and the Office of Naval Research. Dr. Tung has published over thirty articles in engineering journals and conferences. In 2000 and 2002, he was named Outstanding Researcher in Mechanical Engineering.

Chao-Hung S. (Steve) Tung, 2003.
(Mechanical Engineering collection)

Deepak G. Bhat, *BSMetE, University of Poona, India; MTechMetE, IIT Bombay; PhD, University of Southern California; Erma Fitch and Raymond Giffels Professor (2001–present)*

Dr. Deepak Bhat received his bachelor's degree in 1966 in Metallurgy from the University of Poona (India), and in 1968, obtained the master's degree in Metallurgical Engineering from the Indian Institute of Technology, Bombay (India). He was a faculty member in Metallurgy at the College of Engineering, University of Poona, before joining the University of Southern California, Los Angeles, where he earned his a PhD in Materials Science in 1978 for his work on superplasticity in the Zn-Al eutectoid alloy.

Prior to joining the University of Arkansas, Dr. Bhat held a variety of industrial research engineering and management positions in various companies in the United States, spanning a career of over twenty-five years. He

was manager of coatings development at
Stellram/Metalworking Products (a divi-
sion of Allegheny Technologies, Inc.) in
La Vergne, Tennessee. Before that, Dr.
Bhat served as president, UES Arcomac,
Inc., a subsidiary of UES, Inc., and man-
ager of technology marketing and com-
mercialization for UES Materials and
Processes and Surface Modification
Technology Groups, in Dayton, Ohio. At
UES he initiated the development of an
innovative new coating technology for
extending the life and performance of
die-casting dies. Prior to joining UES,
Dr. Bhat was for thirteen years a senior
research scientist and engineering
manager for Advanced Coatings
Development at Valenite, Inc., Madison

Deepak G. Bhat, 2003.
(Mechanical Engineering collection)

Heights, Michigan (a global cutting-tool manufacturer). He also served
as principal process engineer at the San Fernando Laboratories, Pacoima,
California, where he developed a low-temperature, nanocrystalline composite W-W3C coating for improving the erosive and abrasive wear resistance of oil field slurry transport pipeline and machinery. He holds nine
U.S. patents and one Canadian patent on various aspects of cutting-tool
materials technologies and has published more than fifty papers on the
subject.

During his academic career, Dr. Bhat was the recipient of many scholarships, including the Government of India Merit Scholarships
(1957–1961, 1966–1968) and the Daniel Jackling Foundation Fellowship,
University of Southern California (1975). In 1980, Dr. Bhat was selected
as one of the first ASM-IIM Visiting Lecturers to visit various research laboratories and government research centers in India and to present lectures
on advanced CVD coatings for wear applications. In 1993, he was chosen as a member of a delegation by the National Center for Manufacturing
Sciences (Ann Arbor, Michigan) to evaluate advanced coating technologies in the former Soviet Republic of Ukraine. He was invited by the
Government of India in 1998 as a TOKTEN Expert in Surface

Engineering, to present talks and workshops on CVD/PVD techniques. Since 1997, Dr. Bhat has been a visiting lecturer at the Central University of Venezuela (UCV). He has also served in a similar capacity for University of Poona (India), RMIT University (Melbourne, Australia), and Guru Nanak Dev University, Amritsar (India). In 2003, he was appointed as an honorary member of the Board of Curriculum Development in Metallurgy, University of Poona, India. Dr. Bhat is a life member of the Indian Institute of Metals and a member of ASM International, the American Powder Metallurgy Institute, the Society of Manufacturing Engineers, the American Vacuum Society, and the American Society of Engineering Education. He has been listed in the *International Who's Who of Professionals* (1998). He has been active in many technical conferences, having cofounded and chaired the International Conference on Surface Modification Technologies under the auspices of the Metallurgical Society. He has been a consultant to many companies in the United States, Canada, and India.

Min Zou, BSAE, MSAE, Northwestern Polytechnical University (China); MSME, PhD, Georgia Institute of Technology; Assistant Professor (2003–present)

Min Zou joined the Mechanical Engineering Department at the University of Arkansas in August 2003 as an assistant professor. Prior to that, she worked as a staff engineer at Seagate Technology, LLC, from 1999 to 2003 and as a manager at Shanghai Aircraft Research Institute in China from 1991 to 1994. She earned her MS and PhD degrees in mechanical engineering from Georgia Institute of Technology in June 1996 and March 1999, respectively. She also earned her BS and MS degrees in aerospace engineering from Northwestern Polytechnical University (NPU) in China, in June 1988 and January 1991, respectively.

She started her teaching career with Mechanical Engineering Lab I. Her past research included computer hard drive head-disk interface tribology research and development at nano scale, development of a real-time monitoring and control systems for mechanical seals, and aircraft structural analysis. Her current research interests are in the areas of micro/nano tribology and its applications.

She is an active member of the Society of Tribologists and Lubrication Engineers (STLE) and the American Society of Mechanical Engineers

Min Zou, 2003. (Mechanical Engineering collection)

(ASME). She is the recipient of the best paper award (Walter D. Hodson Award) from STLE in 2001. She is also a recipient of the Science and Technology Advancement Award from the Ministry of Aeronautics and Aerospace of the People's Republic of China in 1993.

Appendix

Engineering Mechanics Faculty (1948–1966)

1. Robert Christie Wray, BS, Carnegie Institute of Technology; MSCE, Virginia Polytechnic Institute; MA, University of Arkansas; Professor and Department Head (1948–1966)
2. Francis Eugene Mitchell, BS, Sul Ross State Teachers College; BSCE, University of Texas; Associate Professor (1948–1949)
3. Longley Reed Kirby, BSCE, University of Tennessee; Instructor (1948–1949), Assistant Professor (1949–1956), Professor (1956–1966)
4. William Curtis Clark, BSCE, Alabama Polytechnic Institute; Instructor (1948–1951)
5. John William Crouse, BSCE, University of Arkansas; Instructor (1948–1949)
6. William B. Stiles, BSEE, MSEE, PhD, Iowa State College; Professor (1949–1953)
7. Charles William Crook, BS, Oklahoma A&M College; MS, Brown University; Instructor (1949–1951), Assistant Professor (1951–1956), Associate Professor (1961–1966)
8. Everett Dale Thompson, BSME; Kansas State College; Instructor (1949–1952)
9. Carl Conrad Steyer, BS, Arkansas State College; Instructor (1949–1950)
10. Donald Albert Gilbrech, BSIE, MS, University of Arkansas; PhD, Purdue University; Instructor (1953–1955), Assistant Professor (1955–1958), Associate Professor (1958–1963), Professor (1963–1966)
11. Charles W. Yantis, BSCE, MSCE, A&M College of Texas; Associate Professor (1956–1957)
12. King-Ching Wu, BCC, University of Hong Kong; MSE, University of London; DIC, Imperial College of London; Assistant Professor (1956–1958), Associate Professor (1958–1966)
13. James LaMar Milner, AA, Little Rock Junior College; BS, BSCE, University of Arkansas; Instructor (1956–1957)
14. William Campbell Bryson, BS, United States Naval Academy;

Assistant Professor (1957–1959), Associate Professor (1959–1966)

15. Jimmy Frank Harp, BSIE, University of Arkansas; Instructor (1957–1960)

16. Dana Eugene Daniel, BArchEng, Oklahoma A&M College; Instructor (1957–1958)

17. William Thomas Strickland Jr., BSChE, University of Arkansas; PhD, Rice University; Assistant Professor (1963–1964)

18. Ing-Chang Jong, BSCE, National Taiwan University; MSCE, South Dakota School of Mines and Technology; PhD, Northwestern University; Assistant Professor (1965–1966)

Engineering Science Faculty (1966–1980)

1. Robert Christie Wray, BS, Carnegie Institute of Technology; MSCE, Virginia Polytechnic Institute; MA, University of Arkansas; Professor and Department Head (1966–1975)*

2. Longley Reed Kirby, BSCE, University of Tennessee; Professor (1966–1980)*

3. Donald Albert Gilbrech, BSIE, MS, University of Arkansas; PhD, Purdue University; Professor (1966–1980)*

4. Charles William Crook, BS, Oklahoma A&M College; MS, Brown University; Associate Professor (1966–1980)*

5. King-Ching Wu, BCC, University of Hong Kong; MSE, University of London; DIC, Imperial College of London; Associate Professor (1966–1969)*

6. William Campbell Bryson, BS, United States Naval Academy; Associate Professor (1966–1970)*

7. Ing-Chang Jong, BSCE, National Taiwan University; MSCE, South Dakota School of Mines and Technology; PhD, Northwestern University; Assistant Professor (1966–1969), Associate Professor (1969–1974), Professor (1974–1980)*

8. Donald Wayne Dareing, BSME, MS, University of Missouri; PhD, University of Illinois; Associate Professor (1966–1971), Professor (1971–1980)

9. James R. Kimzey, BSME, University of Arkansas; MSME, PhD, Kansas State University; Assistant Professor (1969–1972), Associate Professor (1972–1980)

10. George D. Combs, BSChE, PhD, University of Arkansas; Associate Professor (1970–1980)
11. Jim H. Akin, BSME, MSME, PhD, University of Texas; Professor and Department Head (1974–1980)
12. James F. Lea, BSME, MSME, University of Arkansas; PhD, Southern Methodist University; Assistant Professor (1975–1980)
13. Carl C. Steyer, BS, Arkansas State University; BSME, MA, University of Arkansas; PhD, University of Texas; Professor (1977–1980)
14. Ali N. Bozatli, BSME, Bosphorus University, Istanbul, Turkey; MSME, Colorado State University; Instructor (1978–1980)
15. Edward P. Clark, BS, United States Naval Academy; MSAE, University of Michigan; Visiting Assistant Professor (1979–1980)

* Also an Engineering Mechanics faculty member.

Notes

Introduction

1. Harrison Hale, *University of Arkansas,* 1871–1948 (Fayetteville: University of Arkansas Alumni Association, 1948), 9, 17, 32.

2. John H. Reynolds and David Y. Thomas, *History of the University of Arkansas* (Fayetteville: University of Arkansas, 1910), 263, 264; Ibid., 62.

3. Monte A. Calvert, *The Mechanical Engineer in America, 1830–1910* (Baltimore, Md.: The Johns Hopkins University Press, 1967), 47–48.

4. Reynolds and Thomas, *History of the University of Arkansas,* 146, 258, 260.

5. University of Arkansas College of Engineering, *College of Engineering Alumni Directory, 1888-1962* (Fayetteville: University of Arkansas, 1963), 56; Hale, *University of Arkansas, 1871–1948,* 65; Reynolds and Thomas, *History of the University of Arkansas,* 264.

6. Hale, *University of Arkansas, 1871–1948,* 73.

Chapter 1: The Formative Years, 1897–1923

1. *University of Arkansas Catalog* (1897–1898), 13–14, 42–43.

2. Ibid., 38–39.

3. Ibid., 155.

4. Ibid., 63.

5. *University of Arkansas Catalog* (1898–1899), 88; *University of Arkansas Catalog* (1902–1903), 6, 92.

6. *University of Arkansas Catalog* (1903–1904), 109.

7. Ibid., 110.

8. *University of Arkansas Catalog* (1917–1918), 163–168.

9. Ibid., 150, 152.

10. Ibid., 115–117.

11. Ibid., 150.

12. *University of Arkansas Catalog* (1897–1898), 12–19.

13. Ibid., 22, 24–26.

14. *Arkansas Engineer* 8 (November 1927), 12; *Arkansas Engineer* 8 (May 1928), 13; *Arkansas Engineer* 15 (May 1936), 7; John H. Reynolds and David Y. Thomas, *History of the University of Arkansas* (Fayetteville: University of Arkansas, 1910), 264.

15. Reynolds and Thomas, *History of the University of Arkansas,* 161; *Arkansas Engineer* 2 (Second Quarter 1922), 11.

16. *Arkansas Engineer* 1 (February 1921), 6; *Arkansas Engineer* 3 (First Quarter 1922–1923), 13; *Arkansas Engineer* 3 (Second Quarter 1922–1923), 15.

17. *Arkansas Engineer* 2 (First Quarter 1921–1922), 7; *Arkansas Engineer* 3 (Second Quarter 1922–1923), 11.

18. *Arkansas Engineer* 3 (Second Quarter 1922–1923), 16.

19. *Arkansas Engineer* 9 (January 1929), 12; *Arkansas Engineer* 11 (March 1932), 4, 10; Bruce Sinclair, *Centennial History of the American Society of Mechanical Engineers, 1880–1980* (Toronto: University of Toronto Press, 1980), 129–131; *Arkansas Engineer*

10 (January 1930), 14; Sinclair, *Centennial History of the American Society of Mechanical Engineers,* 160.

20. Robert A. Leflar, *The First 100 Years: A Centennial History of the University of Arkansas* (Fayetteville: University of Arkansas Foundation, 1972), 289–290, 292–294.

21. *Report of the President* (December 1918), 3–5; Hale, *University of Arkansas, 1871–1948,* 214; Kent R. Brown, *Fayetteville: A Pictorial History* (Norfolk, Va.: Donning, 1982), 66; *University of Arkansas Catalog* (1918–1919), 163–164; *University of Arkansas Catalog* (1925–1926), 138.

22. *Report of the President* (December 1918), 5; *University of Arkansas Catalog* (1919–1920), 136; *University of Arkansas Catalog* (1934–1935), 125.

23. *Arkansas Engineer* 3 (Third Quarter 1922–1923), 9–10.

24. *Arkansas Engineer* 4 (March 1924), 3; *Arkansas Engineer* 2 (First Quarter 1921–1922), 3.

25. Leflar, *The First 100 Years,* 62–63.

26. Ibid., 136–137.

27. Ibid., 299–300, 304–307.

28. Hale, *University of Arkansas,* 1871–1948, 228–230, 312.

29. Ibid.

30. *University of Arkansas Catalog* (1919–1920), 26; *University of Arkansas Catalog* (1923–1924), 21; *University of Arkansas Catalog* (1924–1925), 22; Leflar, *The First 100 Years,* 316–317; *Arkansas Engineer* 18 (November 1938), 5, 22.

31. Brown, *Fayetteville,* 116; Hale, *University of Arkansas, 1871–1948,* 104; *Report of the President* 16 (December 1922), 3.

32. *Cardinal* (1897), 12.

33. *Cardinal* (1904), 16.

34. *Cardinal* (1911), 18; *University of Arkansas Catalog* (1923–1924), 114; *University of Arkansas Engineering Bulletin, 1946–1947,* 24; Cecil Cogburn, interview with author.

Chapter 2: The 1920s through the Depression, 1923–1941

1. *Arkansas Engineer* 4 (December 1923), 11; *Arkansas Engineer* 5 (December 1924), 5; *Arkansas Engineer* 9 (December 1928), 16; *University of Arkansas Catalog* (1930–1931), 162.

2. *University of Arkansas Catalog* (1923–1924), 107–108; *University of Arkansas Catalog* (1923–1924), 117; *Arkansas Engineer* 3 (Second Quarter 1922–1923), 15.

3. *University of Arkansas Catalog* (1923–1924), 107–108.

4. Ibid., 110.

5. *Arkansas Engineer* 5 (March 1925), 5; *University of Arkansas Catalog* (1925–1926), 125; *University of Arkansas Catalog* (1927–1928), 134; *University of Arkansas Catalog* (1929–1930), 145.

6. *University of Arkansas Catalog* (1940–1941), 138, 142, 149–150.

7. Ibid., 139.

8. Ibid., 141.

9. *Arkansas Engineer* 6 (March 1926), 28; *Arkansas Engineer* 7 (November 1926), 11.

10. *Arkansas Engineer* 7 (May 1927), 5, 6.

11. *Engineering News-Record* 99 (November 24, 1927), 827.

12. *Arkansas Engineer* 7 (March 1927), 7; *Arkansas Engineer* 14 (November 1934), 10; *Arkansas Engineer* 13 (March 1934), 16, 20.

13. *Arkansas Engineer* 8 (May 1928), 13.

14. *Arkansas Engineer* 9 (December 1928), 14, 18.

15. Ibid., 18; *Arkansas Engineer* 9 (March 1929), 5–6, 18; *Arkansas Engineer* 10 (March 1930), 12.

16. *Arkansas Engineer* 13 (March 1934), 7.

17. *Arkansas Engineer* 10 (November 1929), 5–6, 14–15.

18. *Arkansas Engineer* 12 (May 1933), 12; *Arkansas Engineer* 13 (November 1933), 12; *Arkansas Engineer* 14 (January 1935), 24; *Arkansas Engineer* 17 (November 1937), 12.

19. *Arkansas Engineer* 16 (November 1936), 12; *Arkansas Engineer* 16 (January 1937), 16.

20. *Arkansas Engineer* 8 (May 1928), 11; Edward L. Gammill, email communication with author.

21. *Arkansas Engineer* 18 (March 1939), 19; Kenneth D. Holloway, email communication with author.

22. *Arkansas Engineer* 11 (February 1932), 10–11; *Arkansas Engineer* 13 (November 1933), 10, 16.

23. *Arkansas Engineer* 11 (March 1932), 4; *Arkansas Engineer* 11 (February 1932), 10.

24. *Arkansas Engineer* 20 (May 1941), 16; *Arkansas Engineer* 31 (November 1951), 22.

25. *Arkansas Engineer* 19 (March 1940), 10; *Arkansas Engineer* 15 (January 1936), 10.

26. *Arkansas Engineer* 18 (March 1939), 9, 22; *Razorback,* 1936, 31.

27. *Arkansas Engineer* 22 (November 1942), 16.

28. Kent R. Brown, *Fayetteville: A Pictorial History* (Norfolk, Va.: Donning, 1982), 110–111, 114, 137, 139.

29. *Arkansas Engineer* 5 (December 1924), 5; *Arkansas Engineer* 6 (November 1925), 5, 14.

30. Cecil Cogburn, interview with author.

31. Cecil Cogburn, interview with author; *Arkansas Engineer* 21 (January 1942), 10.

32. *Arkansas Engineer* 11 (March 1932), 4.

33. *Arkansas Engineer* 25 (November 1946), 8; Edward L. Gammill, email communication with author; David Davison, email communication with author.

34. Alice Ann Cook, daughter, interview with Rick Couvillion.

35. *Arkansas Engineer* 20 (November 1940), 6; *Arkansas Engineer* 25 (November 1946), 10.

36. Cecil Cogburn, interview with author; Thomas Jefferson, correspondence with author.

37. Frank Woolard, email communication with author; David Davison, email communication with author.

Chapter 3: World War II and the Early Postwar Period, 1941–1958

1. *Arkansas Engineer* 28 (January 1949), 32; Ibid., 9; *Arkansas Alumnus* 11 (January 1958), 10.

2. *University of Arkansas Engineering Bulletin* (1946–1947), 10.

3. Ibid., 25.

4. Ibid., 11.

5. *Arkansas Engineer* 26 (November 1947), 17; *Arkansas Engineer* 28 (January 1949), 19, 38, 40–41.

6. *Arkansas Engineer* 30 (May 1951), 21.

7. *Arkansas Engineer* 27 (March 1948), 6.

8. *University of Arkansas Engineering Bulletin* (1957–1958), 9, 44–45.

9. Ibid., 9, 21–22, 24.

10. *University of Arkansas Engineering Bulletin* (1957–1958), 24.

11. *Arkansas Engineer* 21 (November 1941), 13, 25; *Arkansas Engineer* 22 (November 1942), 14, 21–23; *Arkansas Engineer* 23 (April 1944), 6, 21.

12. James H. Crenshaw, email communication with author.

13. Harrison Hale, *University of Arkansas, 1871–1948* (Fayetteville: University of Arkansas Alumni Association, 1948), 132.

14. James H. Crenshaw, email communication with author; Paul A. Kormondy, email communication with author.

15. *Report of the President* 38 (November 1944), 9–10, 14.

16. Ibid., 5–6; Robert A. Leflar, *The First 100 Years: Centennial History of the University of Arkansas* (Fayetteville: University of Arkansas Foundation, 1972), 183, 185; Frederick Rudolph, *The American College and University: A History* (New York: Vintage Books, 1962), 485.

17. *University of Arkansas General Bulletin* (1945–1946), 79; *University of Arkansas General Bulletin* (1946–1947), 81; Martin W. Bates, email communication with author.

18. Martin W. Bates, email communication with author; Paul A. Kormondy, email communication with author.

19. *Arkansas Engineer* 30 (January 1951), 7, 24, 26; *Arkansas Engineer* 31 (January 1952), 12–13.

20. *Arkansas Engineer* 32 (January 1953), 9; *Arkansas Engineer* 35 (November 1955), 30.

21. Barry West, email communication with author; David Davison, email communication with author.

22. *Arkansas Engineer* 21 (January 1942), 10; *Arkansas Engineer* 21 (November 1941), 16; *Razorback* (1942), 256.

23. *Arkansas Engineer* 26 (January 1947) 16; *Arkansas Engineer* 26 (November 1947), 24; *Arkansas Engineer* 28 (November 1948), 38.

24. *Arkansas Engineer* 33 (May 1953), 24.

25. *Report of the President,* 1949–1950, 44 (November 1950), np; *Report of the President,* 1950–1951, 45 (November 1951), 8; *Arkansas Alumnus* 3 (April 1950), 6.

26. *Arkansas Engineer* 25 (November 1946), 12–13; James H. Crenshaw, email communication with author; Cecil Cogburn, interview with author; Edward L. Gammill, email communication with author.

27. *Arkansas Engineer* 29 (January 1950), 15, 28–29; *Arkansas Engineer* 30 (November 1950), 14; *Arkansas Engineer* 31 (November 1951), 35.

28. *Arkansas Engineer* 27 (January 1948), 5; James H. Crenshaw, email communication with author; *Arkansas Engineer* 37 (May 1958), 7.

29. Edward L. Gammill, email communication with author.

30. *University of Arkansas General Bulletin* (1947–1948), 67; Hale, *University of Arkansas,* 1871–1948, 136; Leflar, *The First 100 Years,* 340.

31. Leflar, *The First 100 Years,* 340–341; *Report of the President, 1947–1948,* 42 (November 1948), 18.

32. *Arkansas Engineer* 33 (May 1953), 8; *Arkansas Alumnus* 3:3 (November 1949), 9.

33. *Report of the President, 1947–1948,* 42 (November 1948), 7; Report of the President, 1949–1950, 44 (November 1950), np.

34. Kent R. Brown, *Fayetteville: A Pictorial History* (Norfolk, Va.: Donning, 1982), 142, 149; David P. Davison, email communication with author

35. *Arkansas Alumnus* 11 (January 1958), 6–7.

36. Ibid., 10; Leflar, *The First 100 Years,* 195; *Arkansas Engineer* 36 (March 1957), 6.

37. Lawrence P. Grayson, *The Making of an Engineer: An Illustrated History of Engineering Education in the United States and Canada* (New York: John Wiley and Sons, 1993), 185–186.

38. Leflar, *The First 100 Years,* 275–276, 278–279, 283–285.

39. *Report of the President, 1949–1950,* 44 (November 1950), np; *University of Arkansas Catalog* (1922–1923), 118; *Arkansas Alumnus* 3 (November 1949), 22–23; Joel G. Colton and Stuart W. Bruchey, eds., *Technology, the Economy, and Society: The American Experience* (New York: Columbia University Press, 1987), 12.

40. Leflar, *The First 100 Years,* 340–343.

41. Ibid., 334–335.

42. Hale, *University of Arkansas, 1871–1948,* 135, 226–227.

43. Leflar, *The First 100 Years,* 308–309.

44. Ibid., 308.

45. Ibid., 314–315.

46. Cecil Cogburn, interview with author.

47. *Arkansas Engineer* [46] (May 1968), 5; *Arkansas Engineer* [47] (November 1968), 7; *Arkansas Engineer* [49] (November 1970), 6; Cecil Cogburn, interview with author.

48. Frank Woolard, email communication with author; Charles Ward, email communication with author; Foye Penn, email communication with author.

49. Foye Penn, email communication with author; James Karam, email communication with author.

50. Robert L. Jeske, University of Arkansas Department of Mechanical Engineering personnel files.

Chapter 4: The Jefferson Years, 1958–1969

1. Faculty member in the Department of Engineering Science from 1974 to 1980.

2. *Arkansas Engineer* 38 (November 1958), 7.

3. *Arkansas Engineer* 38 (January 1959), 29.

4. *University of Arkansas Engineering Bulletin* (1958–1959), 25.

5. Ibid., 25.

6. *University of Arkansas Engineering Bulletin* (1967–1968), 20.

7. Ibid., 35.

8. Ibid., 35.

9. *Arkansas Engineer* 39 (January 1960), 29; *Arkansas Engineer* 42 (January 1964), 17.

10. Frank Woolard, email communication with author; *Arkansas Engineer* 44 (May 1966), 18; *Arkansas Engineer* [45] (November 1966), 30; *Arkansas Engineer* [45] (May 1967), 14.

11. *Arkansas Engineer* 40 (May 1961), 23; *Arkansas Engineer* 42 (January 1964), 18; *Arkansas Engineer* 42 (May 1964), 11.

12. Larry Jones, email communication with author; James Karam, email communication with author.

13. *Arkansas Engineer* 42 (January 1964), 26–29.

14. Ibid.

15. Stephen Pile, email communication with author.

16. *President's Report* (1963–1964), 23–24; *President's Report* (1964–1965), 23.

17. *Arkansas Engineer* 39 (November 1959), 11; *President's Report* (1970), 4, 6; *Arkansas Engineer* 41 (March 1962), 7, 34.

18. *Arkansas Engineer* 48 (March 1969), 5.

19. *Arkansas Engineer* 42 (November 1962), 9, 22; *Arkansas Engineer* 43 (January 1964), 27; *Arkansas Alumnus* 14 (February 1961), 10; *Arkansas Alumnus* 15 (February 1962), 10.

20. *Arkansas Alumnus* 15 (February 1962), 10.

21. Ing-Chang Jong, interview with author; *University of Arkansas Engineering Bulletin* (1965–1966), 9.

22. *Arkansas Engineer* [47] (January 1969), 8.

23. *Arkansas Engineer* 39 (March 1960), 9; *Arkansas Engineer* 43 (March 1965), 9, 11–12; *Arkansas Engineer* 43 (May 1965), 22–23; *Arkansas Engineer* [52] (Fall 1973), 8.

24. Robert A. Leflar, *The First 100 Years: Centennial History of the University of Arkansas* (Fayetteville: University of Arkansas Foundation, 1972), 358–359, 362.

25. Ibid., 364–367.

26. Ibid.; Thomas Jefferson, correspondence with author.

27. Leflar, *The First 100 Years,* 204, 206.

28. Ibid., 368–370.

29. Ibid., 295–296.

30. Kent R. Brown, *Fayetteville: A Pictorial History* (Norfolk, Va.: Donning, 1982), 162.

31. Leflar, *The First 100 Years,* 310–311.

32. *Arkansas Engineer* 38 (November 1958), 44; *Arkansas Engineer* 41 (May 1963), 27; *Arkansas Engineer* [47] (November 1968), 6; *Arkansas Engineer* [47] (May 1969), 4.

33. Thomas Jefferson, correspondence with author.

34. Previously taught in the engineering science and engineering mechanics departments.

35. Previously taught in the Engineering Science Department.

36. Henry Hicks Jr., University of Arkansas Department of Mechanical Engineering personnel files.

37. Arthur Bowie, email communication with author.

38. Michael Foote, email communication with author.

39. Helmut Wolf, University of Arkansas Department of Mechanical Engineering personnel files; *Arkansas Engineer* [51] (October 1972), 3.

40. Frank Woolard, email communication with author; Stephen Pile, email communication with author.

41. *Arkansas Engineer* [50] (March 1971), 19.

42. Frank Woolard, email communication with author.

43. Arthur Bowie, email communication with author.

44. Previously taught in the Departments of Engineering Science and Engineering Mechanics.

45. Ing-Chang Jong, email communication with author.

46. Ibid.

47. Ibid.

Chapter 5: The Deaver Years, 1969–1980

1. Faculty member in the Department of Engineering Science from 1969 to 1980.

2. *University of Arkansas Engineering Bulletin* (1969–1970), 36–37.

3. Ibid., 36.

4. Ibid., 37.

5. *University of Arkansas Engineering Catalog* (1980–1981), 49–50.

6. Ibid., 49.

7. Ibid., 50.

8. *Arkansas Engineer* [51] (December 1972), 13; *Arkansas Engineer* [54] (Fall 1975), 14; *Arkansas Engineer* [46] (Spring 1977), 23.

9. Lawrence P. Grayson, *The Making of an Engineer: An Illustrated History of Engineering Education in the United States and Canada* (New York: John Wiley and Sons, 1993), 228.

10. *Arkansas Engineer* 49 (November 1968), 12; *Arkansas Engineer* [52] (December 1971), 10–11.

11. *Arkansas Engineer* 82 (March 1982), 10; *Arkansas Engineer* 84 (March 1984), 4–5; Cecil Cogburn, interview with author.

12. Cecil Cogburn, interview with author.

13. Ibid.; Leon West, email communication with author.

14. *Arkansas Engineer* [48] (November 1969), 6; *Arkansas Engineer* [51] (December 1972), 2; *Arkansas Engineer* [54] (Fall 1975), 3, 7, 14; *Arkansas Engineer* [58] (Fall 1979), 21.

15. *President's Report* (1974–1977), 23; Grayson, 223; *Arkansas Engineer* 51 (December 1971), 5; *Arkansas Engineer* 52 (December 1972), 2, 16; *Arkansas Engineer* 40 (January 1960), 18.

16. *Arkansas Engineer* 55 (Fall 1975), 3; Grayson, *The Making of an Engineer,* 224, 226.

17. *Arkansas Engineer* 82 (January 1982), 9.

18. Grayson, *The Making of an Engineer,* 224, 226, 233.

19. *Arkansas Engineer* 55 (Fall 1975), 3; *President's Report* (1974–1977), 21; *Arkansas Engineer* [55] (Fall 1976), 6.

20. *Arkansas Engineer* [49] (January 1971), 5; *Arkansas Engineer* [55] (Fall 1976), 6.

21. *Arkansas Engineer* [49] (May 1971), 2.

22. *Arkansas Engineer* [49] (May 1971), 2; Ing-Chan Jong, interview with author.

23. Cecil Cogburn, interview with author; Thomas Jefferson, correspondence with author.

24. *Arkansas Engineer* [50] (November 1971), 2, 4.

25. *President's Report* (1971–1972), 30; *President's Report* (March 1, 1974), 10–11.

26. *President's Report* (1971–1972), 6–7.

27. Ibid., 3.

28. *President's Report* (1974–1979), 7; Robert A. Leflar, *The First 100 Years: Centennial History of the University of Arkansas* (Fayetteville: University of Arkansas Foundation, 1972), 317; *Arkansas Engineer* 83 (January 1983), 5.

29. *President's Report* (1974–1979), 7; *President's Report* (1978), 6.
Arkansas Engineer [47] (May 1969), 14; Franklin K. Deaver, University of Arkansas Department of Mechanical Engineering personnel files.

31. *Arkansas Engineer* [47] (November 1968), 7; *Arkansas Engineer* [50] (May 1972), 8; *Arkansas Engineer* [51] (October 1972), 3; *Arkansas Engineer* [95] (Spring 1995), 14; Frank Porbeck, email communication with author.

Chapter 6: Modern Times, 1980–2003

1. Professors Donald A. Gilbrech, Ing-Chang Jong, Longley Kirby and Charles Crook were previously faculty members in the Department of Engineering Science.

2. Ing-Chang Jong, interview with author.

3. *University of Arkansas Engineering Catalog* (1983–1984), 59–60.

4. *University of Arkansas Engineering Catalog* (1981–1982), 50.

5. Ibid., 51.

6. Ing-Chang Jong, interview with author; *University of Arkansas Directory* (1984–1985), 17; *University of Arkansas Engineering Catalog* (1985–1986), 9–10.

7. *University of Arkansas Engineering Catalog* (1985–1986), 61.

8. Ibid., 62.

9. *University of Arkansas Annual Report* (1994–1995), 41.

10. Ibid., 217.

11. *University of Arkansas Engineering Catalog* (1997–1998), 478.

12. *Arkansas Engineer* 85 (Spring 1985), 3.

13. *University of Arkansas Annual Report* (1989–1990), 28; *University of Arkansas Annual Report* (1991–1992), 33; *University of Arkansas Annual Report* (1992–1993), 43.

14. Ing-Chang Jong, interview with author.

15. *University of Arkansas Annual Report* (1994–1995), 41; Lawrence P. Grayson, *The Making of an Engineer: An Illustrated History of Engineering Education in the United States and Canada* (New York: John Wiley and Sons, 1993), 235.

16. *Arkansas Engineer* 81 (March 1981), 8.

17. *Arkansas Engineer* 82 (October 1982), 26–27; *Arkansas Engineer* 83 (October 1983), 18; *Arkansas Engineer* [93] (Spring 1993), 5.

18. *Arkansas Engineer* 83 (October 1983), 18.

19. *Arkansas Engineer* 86 (Summer 1986), 14; *Arkansas Engineer* 86 (Fall 1986), 14.

20. *Arkansas Engineer* [92] (Spring 1992), 5.

21. *University of Arkansas Annual Report* (1985–1986), 17, 50; *University of Arkansas Annual Report* (1986–1987), 54; *University of Arkansas Annual Report* (1987–1988), 29.

22. James Karam, email communication with author.

23. *University of Arkansas Annual Report* (1985–1986), 160–161; *University of Arkansas Annual Report* (1987–1988), 87, 193; *University of Arkansas Annual Report* (1988–1989), 172.

24. *Arkansas Engineer* 81 (March 1981), 14; *Arkansas Engineer* 84 (Fall 1984), np; *Arkansas Engineer* [91] (Spring 1991), 10–11.

25. Rick J. Couvillion, email communication with author.

26. *University of Arkansas Annual Report* (1990–1991), 187–188; *University of Arkansas Annual Report* (1991–1992), 33; Cecil Cogburn, interview with author.

27. *Arkansas Engineer* 86 (Summer 1986), 4; *Arkansas Engineer* 88 (Spring 1988), 8.

28. *President's Report* (1983), 2; *University of Arkansas Annual Report* (1986–1987), 9.

29. Grayson, *The Making of an Engineer,* 226; *Arkansas Engineer* 83 (October 1983), 5.

30. *Arkansas Engineer* 81 (March 1981), 10; *Arkansas Engineer* 80 (October 1980), 14; *Arkansas Engineer* 83 (October 1983), 4.

31. *Arkansas Alumnus* 34 (December 1984), 12; *University of Arkansas Annual Report* (1986–1987), 19; *Arkansas Alumnus* 36 (October 1986), 10; *Arkansas Engineer* 90 (Spring 1990), 4–6.

32. Ing-Chang Jong, interview with author.

33. Rick J. Couvillion, email correspondence with author.

34. Ibid.; *Arkansas Engineer* 86 (Fall 1986), 8.

35. Leon West, email correspondence with author.

36. Ibid.

37. *Arkansas Engineer* [95] (Spring 1995), 15; Larry Roe, email communication with author.

Index